Oxford International Primary

3

W0090553

Science

Workbook

Deborah Roberts

Terry Hudson

Alan Haigh

Geraldine Shaw

Language consultants:

John McMahon

Liz McMahon

OXFORD

OXFORD
UNIVERSITY PRESS

Great Clarendon Street, Oxford, OX2 6DP, United Kingdom

Oxford University Press is a department of the University of Oxford. It furthers the University's objective of excellence in research, scholarship, and education by publishing worldwide. Oxford is a registered trade mark of Oxford University Press in the UK and in certain other countries.

British Library Cataloguing in Publication Data

Data available

ISBN 978-1-382006620

7 9 10 8

Paper used in the production of this book is a natural, recyclable product made from wood grown in sustainable forests. The manufacturing process conforms to the environmental regulations of the country of origin.

Printed in China by Golden Cup

Acknowledgements

The publisher and authors would like to thank the following for permission to use photographs and other copyright material:

Cover: Artwork by Blindsalida. Photos: **p17:** wundervisuals/E+/Getty Images; **p19:** Pukao/Shutterstock; **p28:** World History Archive/Alamy Stock Photo; **p34:** Blue Jean Images/Alamy Stock Photo; **p38(a):** Auris/Dreamstime.com; **p38(b):** B art/Shutterstock; **p38(c):** Tezzstock/Dreamstime.com; **p38(d):** Borja Andreu/Shutterstock; **p38(e):** Marekuliasz/Shutterstock; **p60:** thirboy/istockphoto; **p105(a):** R S Vivek/Dreamstime.com; **p105(b):** Skydive Erick/Shutterstock; **p105(c):** Open FIlms/Shutterstock; **p105(d):** Levent Konuk/Shutterstock; **p105(e):** Ilukee/Alamy Stock Photo; **p119(a):** Omika/Adobe Stock; **p119(b):** oTTo-supertramp/Shutterstock; **p119(c):** Zuzule/Shutterstock; **p119(d):** Mogens Trolle/Shutterstock; **p119(e):** Massimhokuto/Adobe Stock; **p119(f):** Alex Churi-lov/Shutterstock.

Artwork by Q2A Media.

Every effort has been made to contact copyright holders of material reproduced in this book. Any omissions will be rectified in subsequent printings if notice is given to the publisher.

Contents

How to Use this Book

The Workbook for *Oxford International Primary Science* supports the Student Book that children are using in their science lessons for this year.

The Student Book includes some pair, group and whole-class activities, hands-on tasks and write-in tasks to test students' understanding and help them learn. It is important to extend these tasks. This Workbook enables students to build on what they have learned in the Student Book to develop a secure understanding of scientific concepts.

Encouraging students to think about and apply their growing skills and knowledge helps them consolidate their understanding and work scientifically. This helps with confidence. Students also have opportunities to see that science is relevant all around them – both inside and outside the classroom.

Students may find it useful to complete an investigation planning form. This sets out all the stages of the investigation. A proforma is provided in the Teacher's Guide. Find out more at:

www.oxfordprimary.com/international-science

Structure of the book

This Workbook is divided into five units plus a Support for Teachers and Parents section and a Quiz:

Support for Teachers and Parents

Unit 1 Light and Dark

Unit 2 Looking at Rocks and Soil

Unit 3 Flowering Plants

Unit 4 Introducing Forces and Magnets

Unit 5 Exploring Health, Skeletons and Muscles

Quiz Yourself

What you will find in each unit

There are four types of lessons:

Key words and introduction lessons encourage students to read, spell and use the scientific vocabulary in the unit.

Activities build on the work in the Student Book. These help with developing language skills, developing scientific enquiry skills, applying mathematical knowledge, and securing understanding rather than just recall. The Support for Teachers and Parents notes on pages 6–13 give you advice on how to help students with each activity.

What have I learned encourages students to talk about what they have learned, reflect on what went well and revisit any areas they need to check. This encourages a growth mindset.

Investigate like a scientist enables students to apply what they have learned in practical contexts.

What you will find in the lessons

Icons show the nature of each task:

Discuss: Students are encouraged to discuss and communicate scientific ideas and approaches. They can work in pairs or small groups for discussion tasks.

Investigate: Students are encouraged to plan, ask questions and record results for each investigation. They are asked to observe closely, make predictions and compare their results with others. Sometimes you will use different equipment, which is available in school. You may also ask students to carry out a test in a different way, to make sure they are safe.

Language support: This icon highlights activities that provide language support through writing frames or word boxes. Students are encouraged to write, read and record short answers.

Hints and tips: Students are encouraged to think about tips to make investigations safer or more effective.

Stretch zone: Students are encouraged to extend their understanding.

Mindful moments: Students are encouraged to think about and reflect on what they have learned. This supports students' well-being.

What went well: Students are encouraged to talk about what went well in each module to secure their understanding.

Student Book

Throughout the Workbook, you will find links to the Student Book. Students can refer to information in the Student Book to help them complete activities.

Teacher's Guide

The Teacher's Guide that accompanies this book provides lesson notes and answers for each page.

Support for Teachers and Parents

1 Light and Dark

What students will learn

This unit helps students to understand more about light and dark. They explore sources of light and look at what happens when light shines on materials. Students learn about reflection and objects that either block light or let light through. They then investigate shadows. Students will:

- find out about different light sources
- discover that darkness is when there is no light
- notice that light is reflected from some surfaces
- understand how to protect eyes from the Sun
- find out how shadows are made
- explore and find patterns in shadows.

> #### Key words
> dark, light, pattern, protect, reflect, shadow, Sun, torch

Scientific enquiry skills

This unit helps students to develop and practise the following scientific enquiry skills.

Scientific enquiry skill	Page
Use a range of equipment	17, 18, 19, 21, 27
Make careful observations	14, 16, 18, 21, 27
Take accurate measurements	25
Identify differences, similarities or change	21, 22, 27
Record data in a variety of ways	19, 24, 25, 27
Plan and carry out fair tests	24, 25
Group/classify	24, 27
Use secondary sources of evidence to support ideas	23
Communicate findings and conclusions in a range of ways	16, 18, 21, 23, 27

Ways to help

- Encourage students to use the key words and display them in the room.
- Set out a range of objects that are opaque or transparent so students can explore them.
- Ask students questions about their experiences of being in the light and the dark.
- Ask students to think about when they have seen and used shadows.
- Play games by asking students to suggest when they have seen large and small shadows.
- Arrange to be able to darken a room so students can compare light and dark.

Helping with activities

The following guidance gives you advice on how to help students with each activity.

Sources of light
Arrange a display of different sources of light, such as torches, lamps and candles.

The Sun
Pre-cut the holes in the envelopes and make sure the mirrors fit easily into them.

Does a mirror reflect light?
Arrange some dull and shiny objects around the room so that students can find a range of reflective and non-reflective materials.

Reflecting light
Darken the room as much as possible so the reflections from the torch are more dramatic.

Investigating darkness
Make sure everyone in the room is very quiet when someone is blindfolded so they can hear outside noise they might otherwise not notice.

Make a dark box
Include a lit torch in the list of objects so students see that some things give out (emit) light and others do not.

Light sources
Encourage students to imagine what their life would be like if there were no light sources after the Sun had gone down.

Light to see
Ensure that students understand that natural objects or sources exist in nature and have not been made by people.

Does light travel through all objects?
Darken the room slightly so that any shadows cast are more visible and include transparent and opaque objects.

How do shadows change?
Help students to stick a long ruler or tape measure to the table so they have a fixed measuring point.

2 Looking at Rocks and Soil

What students will learn

This unit helps students to understand more about rocks and fossils. They will investigate some common rocks, including soil, and discuss their uses. Students will learn about the properties of rocks that make them useful. They will then explore how fossils are formed. Students will:

- name some types of rock
- compare and group rocks by what they look like and their properties
- learn how fossils are formed
- investigate different types of soil.

Key words

crystal, fossil, grain, group, property, rock, sand, soil, stone

Scientific enquiry skills

This unit helps students to develop and practise the following scientific enquiry skills.

Scientific enquiry skill	Page
Use a range of equipment	32, 35, 39, 41, 43
Make careful observations	28, 29, 30, 32, 33, 36, 37, 38, 39, 41, 43, 45
Take accurate measurements	43, 45
Identify differences, similarities or change	30, 31, 34, 37, 41, 45
Record data in a variety of ways	30, 31, 32, 34, 36, 37, 39, 43, 45
Plan and carry out fair tests	39, 41, 43, 45
Group/classify	31, 32, 33, 34, 41, 42
Use secondary sources of evidence to support ideas	29, 30, 36
Communicate findings and conclusions in a range of ways	29, 30, 34, 35, 36, 37, 39, 41, 45

Ways to help

- Set out a display of different rocks and fossils so students can handle them.
- Encourage students to observe rocks on their way to and from school.
- Ask students questions about the rocks they have seen and used.
- Ask students to write the key words onto pieces of card and display them in your room.
- Allow students to identify rocks and fossils in books and on the internet, and download pictures.
- Ensure that students always link what a rock is used for with its properties.

Helping with activities

The following guidance gives you advice on how to help students with each activity.

Finding rocks
Point out that not all rocks are solid and hard. Sand and mud are examples of rocks.

Make a model Earth
Explain the difference between the inner core, outer core, mantle and crust.

Identifying rocks
Explain that an identification key works by asking one question after another to narrow down on the rock type.

Label the rock cycle
Explain what igneous, sedimentary and metamorphic rocks look like. Show students some visual examples.

Fossil field trip
Remind students that any evidence of living things in the past is a called a fossil – including footprints.

Fossil hunt
Demonstrate how to lay the string across the trays and fix them in place with tape. Students will need four lengths of string across the tray and four lengths down the tray.

Building materials
Download or project some examples of famous buildings around the world to show different building materials.

Building materials survey
Explain how to take a tally instead of using numbers. Count out three pencils one after the other and write I then II and then III.

Using rocks
Help students to see the link between properties and uses by asking why a soft rock would be useless for steps into a building.

Testing rock hardness
Ensure students can test a range of rocks from soft to hard – and point out that any pieces that break away may be small so they will need to look carefully.

What is in soil?
Explain that the soil is separated into its different parts because heavy parts will sink and lighter parts will float.

Investigating different soils
Ask students to look back at the diagram on the previous activity page for clues as to what they may observe.

Different types of soil
Display the names of the different soil types on a wall so students become familiar with the terms. Loam is best for growing and students could remember this from the acronym **L**ots **O**f **A**mazing **M**aterials.

Soil investigation
Encourage students to look carefully at the soils to note colour and how much sand or clay is present before making their predictions.

3 Flowering Plants

What students will learn

This unit helps students to understand more about plants. They will study the structure of plants and investigate what is needed for healthy plant growth. Students will consider water, light, air and nutrients from soil and place the stages of plant growth into a life cycle. Students will:

- explore plant roots, leaves, stems, trunks and flowers
- find out that plants need water, light, air, nutrients from soil, and space to grow
- investigate how water is taken in and moves through plants
- explore how temperature changes the way plants grow
- identify and describe parts of flowering plants
- find out how flowers are involved in the life cycle of flowering plants.

> **Key words**
> dispersal, flower, growth, leaf, nutrient, pollen, pollination, reproduce, root, seed, stem, transport, trunk, water

Scientific enquiry skills

This unit helps students to develop and practise the following scientific enquiry skills.

Scientific enquiry skill	Page
Use a range of equipment	48, 52, 53, 56, 58, 59, 62, 64, 65, 69
Make careful observations	49, 51, 52, 53, 54, 56, 58, 60, 61, 62, 64, 69
Take accurate measurements	53, 54, 56, 58, 62, 64
Identify differences, similarities or change	49, 52, 53, 56, 58, 59, 60, 61, 62, 63, 64
Record data in a variety of ways	48, 49, 52, 53, 54, 56, 58, 61, 62, 63, 64, 67
Plan and carry out fair tests	52, 53, 54, 56, 58, 64, 69
Group/classify	60, 62
Use secondary sources of evidence to support ideas	49, 59, 62
Communicate findings and conclusions in a range of ways	48, 49, 51, 52, 53, 54, 56, 58, 62, 63, 67, 69, 71

Ways to help

- Encourage students to use the key words and display them in your room.
- Allow students to plant seeds before the start of the topic so they have a range of small plants to investigate.
- Identify a local garden or park where students can carry out visits to observe plants.
- Ask students questions about the plants they have seen and used locally.
- Ask students to think about when they have seen unhealthy plants and ask them where this was and why.
- Have numerous different potted plants in the room so students can compare a range of flowering plants.

Helping with activities

The following guidance gives you advice on how to help students with each activity.

Make a model of a plant

Choose a very wide range of materials so students have a good choice when building their model. Explain that function means the job something does.

Flowering plant exhibition

Make sure students remove any labels pointing to the parts of a flower before asking visitors to identify them.

Flowering plant wordsearch

Remind students that words can be written upwards, downwards, across and diagonally in a wordsearch.

Plant rescue

Remind students that unhealthy plants will not be able to grow large green leaves and strong stems.

Do plants need water to grow?

Remind students to make their investigation a fair test by changing only one thing.

Measuring plants

Ask students to look carefully at the ruler or tape measure to ensure their measurements are accurate and to repeat them three times.

Plants and light investigation

Encourage students to use their observation skills carefully and record details such as colour and number of leaves.

Photosynthesis competition

Explain that only letters from the word 'photosynthesis' can be used to make words in the competition.

Grow a bean seed

Remind students to keep the bean seed damp so it does not dry out during the investigation.

How water moves through plants

Emphasise that water moves up the plant from the roots, and make sure students do not think that water enters through the leaves and 'falls' down inside the plant.

Do leaves affect how a plant transports water?

Explain that plants lose water through holes in the leaves. The more leaves there are, the more water will be drawn up the plant.

Stems for support

Help students to fix heavy pieces of clay to represent leaves so the single strand of spaghetti or straw bends and breaks.

The power of roots

Explain that, given enough time, roots can push through brickwork and even tiny gaps in concrete.

Plant expert's report

Explain that roots need to spread out in soil to take in enough water, and if they are in a small pot this cannot happen.

Tree survey

Remind students that the circumference of a tree is the distance around the trunk. Point out the word is similar to circle.

Unfair competition

Encourage students to use all of their prior knowledge about plant growth to help with their prediction.

Seedling investigation

Make sure students understand that seeds do not need light to germinate but seedlings do need light.

Make your own greenhouse

Download some pictures of model greenhouses, such as ones made from plastic bottles, to give students some clues if they cannot think of a design.

The parts of a flower

Make, or allow students to make, large poster-sized labelled diagrams of flowers so they are constantly in view to help learning and recall.

Identify the parts of the flower

Explain that this activity draws together learning from the investigation in the Student Book and students' prior knowledge of the functions of the parts of the flower.

Fertilisation

Emphasise that fertilisation only refers to the moment the pollen joins with the ovule. The rest of the process is there to help this to happen.

Investigating seed dispersal

Remind students that the independent variable is the variable they decide to change. The dependent variable is what they will be observing or measuring.

4 Introducing Forces and Magnets

What students will learn

This unit helps students to understand more about forces. They will explore a range of different pushes and pulls and learn about the ways in which forces can make objects move faster or slower, or change direction. Students will study friction as a force that slows objects. They will then explore magnetism and learn that magnets can attract and repel each other. Students will:

- revise that pushes and pulls are examples of forces
- explore how forces can make objects start or stop moving

- explore how forces including friction can make objects move faster or slower or change direction
- understand that some forces need contact between two objects but magnetic forces act at a distance
- find out that magnets can attract and repel each other
- find out that magnets attract some materials, but not others
- explore the two poles of a magnet.

> ### Key words
> attract, contact force, force, friction, magnet, non-contact force, pole, pull, push, repel

Scientific enquiry skills

This unit helps students to develop and practise the following scientific enquiry skills.

Scientific enquiry skill	Page
Use a range of equipment	76, 77, 78, 79, 80, 81, 82, 83, 85, 88, 89, 90, 92, 93, 95, 96, 97, 98, 99, 101
Make careful observations	74, 75, 76, 77, 78, 79, 80, 81, 82, 83, 85, 88, 89, 92, 95, 96, 98, 99, 101
Take accurate measurements	76, 77, 80, 81, 82, 83, 85, 86, 98, 99, 101
Identify differences, similarities or change	76, 78, 79, 80, 81, 83, 85, 88, 96, 98, 99, 101
Record data in a variety of ways	76, 77, 78, 80, 81, 82, 83, 85, 86, 88, 92, 95, 98, 99, 101
Plan and carry out fair tests	76, 80, 81, 82, 83, 85, 95, 98, 99, 101
Group/classify	76, 79, 85, 88, 96
Communicate findings and conclusions in a range of ways	72, 74, 75, 76, 77, 81, 82, 83, 85, 93, 95, 97, 98, 99, 101

Ways to help

- Create push and pull labels and fix them to objects that are pushed or pulled.
- Encourage students to say the words 'push' and 'pull' when they open or close doors.

- Allow students to play with objects such as toys and balls to explore forces in everyday contexts.
- Ask students questions about the forces they use every day.
- Ask students to think about what would happen if objects didn't move or couldn't slow down if they were moving.
- Arrange a display of magnets and show some of their everyday uses – such as fridge magnets.

Helping with activities

The following guidance gives you advice on how to help students with each activity.

Find the forces
Allow students time to spot all of the everyday examples of forces and then encourage them to talk about how they have used forces in this way.

Showing the direction of a force
Explain that arrows are often used to show the direction of forces and that they should be shown starting from where the force starts.

Make a forcemeter
Explain that a large force will stretch the forcemeter further than a smaller force.

Using a forcemeter
Explain that forces are measured in units called Newtons and that this can be written as N. At this stage, students just need to know that a force of 10N is twice as much as a force of 5N.

Make your own play dough
Encourage students to shape the dough in as many ways as they can. Some may forget to twist and stretch the dough, for example.

Using forces to change the shape of modelling clay
Emphasise the measuring and comparison skills being used as students explore the shapes of the clay.

Bouncing ball investigation
Encourage students to link the force they apply to the ball with its speed to the wall and its return after bouncing.

Measuring distance
Display a range of distance measuring devices such as a 30-cm ruler, a metre stick, and various lengths and styles of tape measures.

Vehicles travel further on some surfaces
Explain that this is a planning activity to help students to predict and control an investigation to make it a fair test.

Surface investigation

Allow students to use forcemeters if possible but, if not, they can use their own based on elastic bands. Encourage them to measure the lengths of the elastic band to estimate the strengths of the forces.

Which shoe to use?

Remind students that when there is a lot of friction between surfaces the surfaces will not slide past each other very well.

Which shoe has the best grip?

Explain that the larger the force of friction the more force will be needed to move the shoe along the surface.

Rolling a ball

Encourage students to think about the link between the amount of force applied to an object and how quickly or slowly it moves.

Forces and moving objects

Ask students to read the sentences out loud and say the words in the box before they try to fill in the gaps.

Which objects are magnetic?

Make sure you have a range of different steel objects around the room to act as magnetic materials for students to test.

How can we identify the poles of magnets?

Remind students that the North-seeking pole of a magnet will repel the North-seeking pole of another magnet but will attract a South-seeking pole.

Make a compass

Allow students to see and use some navigation compasses so they can see how magnets are used to find direction.

All about magnets

Allow students to look back through their books if they cannot find a correct word. This is a useful skill to develop.

Buried treasure

Point out that magnets can detect and attract magnetic materials through other substances if the layers are not too thick.

Design a recycling plant

Allow students to test their magnets on some of the cleaned waste objects – especially aluminium and steel cans.

Why does the Earth have a North Pole and a South Pole?

Encourage students to fill in as many of the diagram labels as they can from memory and only check with the Student Book if they are definitely stuck.

How do magnets react together?

Encourage students to discuss the investigation and how they will approach it, as this will help to develop their enquiry and investigative skills.

Magic magnets

Demonstrate how you can make a paperclip look as if it is moving on its own by moving it with a hidden magnet.

Floating paperclips

Explain that a floating magnet shows that magnetic force acts at a distance – the magnet and paperclip do not need to be touching.

Making and testing an electromagnet

Demonstrate how to coil the wires around the nail so the coils are compact and neat.

More about electromagnets

Show students how to make the electromagnet with an air coil by coiling wire around a nail first and then removing the nail.

5 Exploring Health, Skeletons and Muscles

What students will learn

This unit helps students to understand more about health, skeletons and muscles. They will explore a range of different life processes. They will learn about the similarities between skeletons and muscles in humans and animals. They will also explore how medicines are used. Students will:

- review the life processes of nutrition, movement, growth and reproduction
- find out that animals and humans need to get their nutrition from what they eat
- find out that humans and some animals have bony skeletons
- discover that animals with skeletons have muscles attached to the bones
- understand how we use medicines.

Key words

bone, diet, exercise, food, healthy, infectious disease, medicine, movement, muscle, nutrition, reproduce, skeleton, vaccination

Scientific enquiry skills

This unit helps students to develop and practise the following scientific enquiry skills.

Scientific enquiry skill	Page
Use a range of equipment	110, 111, 115, 116, 120, 125, 129, 135
Make careful observations	105, 110, 111, 115, 117, 118, 119, 120, 124, 125, 127, 128
Take accurate measurements	110, 111, 115, 120
Identify differences, similarities or change	106, 115, 120, 125
Record data in a variety of ways	110, 115, 120, 121, 125, 131
Plan and carry out fair tests	110, 111, 115, 120
Group/classify	106, 113, 125
Use secondary sources of evidence to support ideas	104, 106, 108, 113, 114, 125
Communicate findings and conclusions in a range of ways	104, 107, 109, 115, 116, 120, 125, 129, 132, 135

Ways to help

- Ask students to share their ideas about life processes to find out what they already understand.
- Allow students to look up x-ray photographs in books and on the internet.
- Display photographs of healthy and unhealthy meals to encourage discussion.
- Ask students to think about what they eat and why they need to eat it.
- Display some health leaflets from a local medical centre.

Helping with activities

The following guidance gives you advice on how to help students with each activity.

Designing a poster
Have some exemplar posters in your room to show students what eye-catching posters look like.

Identifying life processes
Encourage students to tick off each word in the box as they use it.

How much should we eat?
Explain that the figures in the table are averages for males and females, and that some females will use more energy than some males.

Healthy eating
Remind students that advice from health experts is that people should eat only a small amount of fats and sugars.

Signs and symptoms
Students could act out some of the symptoms to help understanding (e.g. sneezing, chills, stomach pain).

Lifestyle diseases
Explain to students that food labels list amounts of vitamins and minerals contained. Make sure they understand the units of measurements used on the food labels.

Looking after water
Help students to set a tap dripping by demonstrating so they see one drop hit the sink just as another drip starts falling from the tap.

Filtering water
Remind students of other filters they have seen, such as coffee filters, tea bags and air filters in vacuum cleaners.

Food for a long walk
Ask students to review the different nutrients found in food and to identify those that provide the most energy.

Different people need different diets
Encourage students to think about how tired they would be if they did each activity for a few hours; this will help them to realise which need the most energy.

Energy and exercise
Explain that muscles need energy to help them to work and this energy needs to be provided by eating foods rich in energy.

Heart rates
Finding a pulse is not always easy. Demonstrate how to feel the pulse in the wrist and remind students to use their finger and never the thumb, which has its own small pulse.

Make a skeleton
Explain that students need to make their own copy of the bones but much larger, otherwise their puppet will be far too small.

Label the skeleton
Allow students to look back at the labelled picture in the Student Book, but only after they have tried to remember each label first.

Comparing animal skeletons
Explain that vertebrates can have very similar skeletons as the bones have similar jobs to do, but there are important differences such as length and shape of some bones.

Matching animals to their skeletons
Explain that the shape of the skeleton helps to make the overall shape of the animal, so students could draw around each skeleton to get the shape of each animal.

Growing bigger
Remind students to measure from the same starting point every time they compare the length of a bone so it is a fair test.

Do all bones grow at the same rate?
Help students to work out the average (mean) length of each part of the skeleton by working out the first one with them.

Protecting your organs
Explain that students should add the organs in the correct place and then label the parts of the skeleton that protects each one.

The role of the ribcage
Remind students to use words from the box only to fill in the gaps, and to use each word once only.

Seeing through bodies
You can access and download examples of x-ray photographs from the internet and display them in the room so students see more examples.

Invertebrate survey
Encourage students to use their observation skills carefully before deciding on a yes or no answer when using keys, otherwise they could move down the wrong branch of the key.

Important muscles
Ask students to move their arms and legs as they feel some of the muscles moving.

Using muscles
Students could act out the activities to learn which muscles are being moved.

Look, cover, label, check
Use this strategy for any diagram that students need to learn, as it is a powerful way to help them remember details.

Muscle pairs
Explain that the place that the lower and upper arm meets needs to be free to move. It is the elbow in their body.

Medicines
Make sure that students understand that each word in the box can be used more than once, otherwise they will find some gaps they cannot fill.

Allergies
Allow students to add allergies that are not in the table, if they are mentioned.

Health leaflet
Display a range of leaflets from a local health centre so students have ideas about the content but also how leaflets can be folded.

Medicines crossword
Explain that the questions can be answered in any order so students could start with the ones they find easier and then have letters in place to give them clues about the harder answers.

Key words

One of these rooms is dark.

One of these rooms is light.

dark
light

dark
light

Colour in the correct word under each picture.

 **Stretch zone**

Why do we need light? Think of one reason.

When is dark useful? Think of one example.

Introduction

Light

1 Find the key words in the wordsearch.

> dark light pattern protect
> reflect shadow Sun

m	l	a	a	p	c	f	d
d	b	m	h	r	t	b	u
s	h	a	d	o	w	e	c
g	c	l	e	t	g	f	s
h	a	s	w	e	d	b	u
b	c	f	d	c	j	c	n
l	i	g	h	t	b	n	e
j	b	y	h	f	d	a	s
p	a	t	t	e	r	n	d
c	w	e	b	g	c	h	a
r	e	f	l	e	c	t	r
x	a	f	i	a	b	c	k

2 Check the key words in your Student Book.

Which key word is missing? _____

3 Draw a picture to show what this word is.

Where does light come from?

Sources of light

There are many sources of light around us.

1 Look around you.

Draw a picture of a source of light that you can see.

```
..........................................................................
:                                                                        :
:                                                                        :
:                                                                        :
:                                                                        :
:                                                                        :
:                                                                        :
:                                                                        :
:                                                                        :
:                                                                        :
:                                                                        :
..........................................................................
```

2 Look outside.

Draw a picture of a source of light that you can see.

```
..........................................................................
:                                                                        :
:                                                                        :
:                                                                        :
:                                                                        :
:                                                                        :
:                                                                        :
:                                                                        :
:                                                                        :
:                                                                        :
..........................................................................
```

Stretch zone

Label each drawing.

The Sun

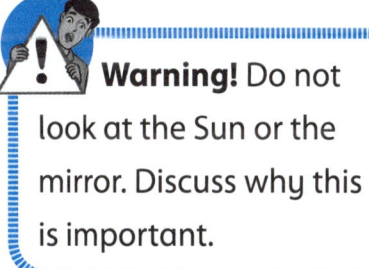**Warning!** It is dangerous to look at the Sun. It is so bright that it could damage your eyes.

There are ways of seeing the Sun – but not directly.

1   Cut a hole in the front of an envelope. This should be 1 to 1.5 centimetres across.

2   Put a small mirror into the envelope.

3   Hold up the envelope so the mirror is pointing towards the Sun.

4   Turn the envelope so the mirror shines an image of the Sun onto a wall.

Warning! Do not look at the Sun or the mirror. Discuss why this is important.

5   Move towards and away from the wall until you get a clear image of the Sun.

Is a mirror a source of light?

Does a mirror reflect light?

You are going to test whether a mirror reflects light.

1 Make the room less bright. Stand near a torch or small lamp.

2 Look around the room. What can you see?

3 Do any of the objects reflect light from the torch or lamp? List them below

4 Take a small mirror. Try to reflect light from the torch or lamp. Light up any dark corners of the room.

5 An example where light is reflected by a mirror is

_____.

Draw your example below.

Reflecting light

You are going to investigate whether all smooth objects reflect light.

You will need: a torch or a table lamp.

1 Find objects around the room that feel smooth.

2 Record the objects in the table below. You could take photographs or draw the objects.

3 Shine a source of light onto each smooth object. Does it reflect the light?

Draw a tick ✓ in the table if it does reflect light.

One example has been done for you.

Name of smooth object	Picture of the object	Does it reflect light?
plate		✓

What is darkness?

Investigating darkness

In darkness we use our other senses to help us move around safely.

Hearing helps us to get to places safely when we cannot see.

How could you test this?

a Wear a blindfold. Sit very still and quiet.

b Think about all the sounds you can hear.

c Take off the blindfold. What can you hear now?

1 When could you hear the most noises? Why?

_____

_____

2 Did your investigation show that our hearing helps us when we cannot see?

Circle your answer.   **yes**   **no**

Think about what you have learned.

Make a dark box

You are going to investigate which objects give out light.

You will need: a cardboard box, a selection of objects to place in the box.

1 Cut a small round hole in the top of the cardboard box.

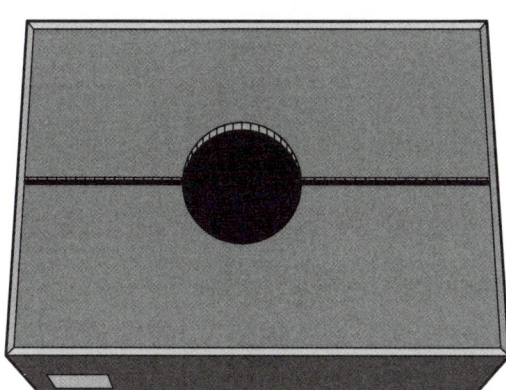

2 Carefully hold the box so it doesn't move around. Look in through the hole.

- What can you see?
- Describe to another person what it looks like in the box.

3 Now close your eyes and wait for your partner to place something in the box.

- Look in the hole.
- Describe what you can see to your partner.

4 Open the box. Have another look at the object in the light.

- Describe what the object looks like now.
- How is this different from how it looked in the closed box?

5 Repeat this with different objects.

 Stretch zone

What does this investigation prove?

_____

_____

We need light to see things

Light sources

When there is no light source, it is dark.

1 How can you still see things at night? Draw a picture of a source of light that you use at night.

2 Think of other sources of light that you could use.

a Which source of light would you use to read a book at night?

_____

b Which sources of light would you use during the day?

_____

_____

_____

Stretch zone

Why would you use these sources of light at different times of the day?

3 Which light source should you never look at? _____

4 What can you wear to protect your eyes when you are outside in bright sunlight?

_____

Light to see

People make things that give out light.

 Can you switch a light on and off?

- If you can switch a light on and off, it is usually made by humans. We say this is **human-made**.

- If you can't switch off a light, it is usually **natural**.

1 Look at the pictures.

Sun   lamp   active volcano   torch

a Circle the natural sources of light in green.

b Circle the human-made sources of light in blue.

2 Which source of light would you choose to use to read this page?

_____

3 Which source of light would hurt your eyes if you looked at it without protection?

_____

Investigating shadows

Does light travel through all objects?

 You are going to investigate whether light can travel through different objects.

You will need: a torch or a table lamp, a selection of objects.

1 Shine your light source onto the first object. Does the light travel through the object or does the object stop the light?

2 Now test the other objects in the same way.

3 Complete the table below. One example has been done for you.

Object	Does light travel through the object?
book	no

4 When light does not travel through the object, what can you see?

Complete the sentence. Choose the correct answer from the box.

I can see _____.

> dark   light   a shadow

How do shadows change?

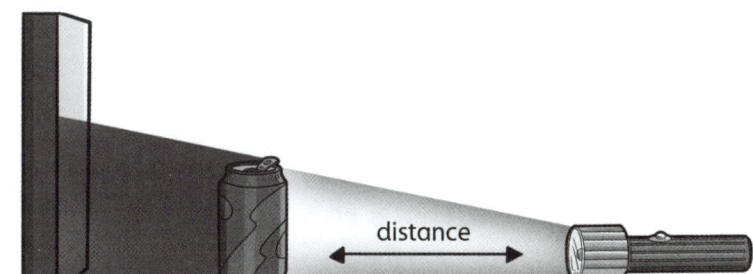

white screen   can   torch

You will investigate what happens to the size of a shadow when an object moves. The room will need to be dark.

1 Look at the diagram. Set up a torch, an object and a screen to match the diagram.

2 Start with the object 20 centimetres from the torch.

3 Then measure the height of the shadow on the screen.

4 Try moving the object 40, 60, 80 and 100 centimetres from the torch. Measure the height of the shadow each time.

5 Record your findings in the table below.

Distance of the object from the torch (cm)	Height of the shadow on the screen (cm)
20	
40	
60	
80	
100	

6 Is there a pattern in the way that the shadow changed? What is it?

Think about what you have learned in this investigation.

What I have learned about light and dark

 What went well

1 Think about what you have learned.

2 Talk to a friend about something that went well in this unit.

3 Tick ✓ the boxes to rate yourself.

I can identify different light sources including the Sun.	That's easy. ⬜ That's challenging. ⬜	Pages 16–19
I know that darkness is the absence of light.	That's easy. ⬜ That's challenging. ⬜	Pages 20–23
I can identify shadows.	That's easy. ⬜ That's challenging. ⬜	Pages 24–25

If you want to know more or need to check, go back to the pages in your Student Book.

Investigate like a scientist

1 Light survey

a   Find different sources of light around you. Draw three sources of light.

b

Look at the picture. Circle the light sources.

● Which are made by humans?

● Which are natural sources of light?

2 Matching shadows

Find objects that you know will make good shadows.

a   Use a torch to make each shadow.

b   Carefully draw around the shadow on a clean piece of white paper.

c   Display the shadow drawings and the objects. Ask another person to match the shadow drawings to the objects.

d   Did they match them all?

2 Looking at Rocks and Soil

Key words

1 Find some of the key words in this unit in the wordsearch below.

r	x	m	l	s
s	o	i	l	t
a	r	c	o	o
n	t	o	k	n
d	t	l	b	e
y	x	r	t	b

rock sandy soil stone

Hint: One word is left to right. Two words are top to bottom. One word is diagonal.

2 Complete the sentences using the words in the wordsearch.

A gardener planted a small tree in some __ __ __ __.

A beach feels very __ __ __ __ __ under your feet.

Many buildings are made from smooth blocks of __ __ __ __ __.

Statues can be carved out of large pieces of __ __ __ __.

Introduction

Describing rocks

1 Find an interesting rock to look at.

- Look closely at your rock.
- What colour is it? Does it have more than one colour?
- What does it feel like? Is it rough or smooth?

2 Use the words in the box to help you describe your rock. You can use your own words too.

> black   brown   crumbly   dirty   grey   hard
> pink   rough   shiny   smooth   white   yellow

The colour of my rock is _____.

My rock is _____ and _____.

3 Draw a picture of your rock.

What are rocks?

Finding rocks

Look at the picture of the beach.

1 Circle any examples of rocks that you can see.

2 Write down one way that rocks are being used.

_____

3 What are the smooth, rounded stones called?

_____

Stretch zone

Find out what sand is made from. Draw some pictures to show what you found out.

Make a model Earth

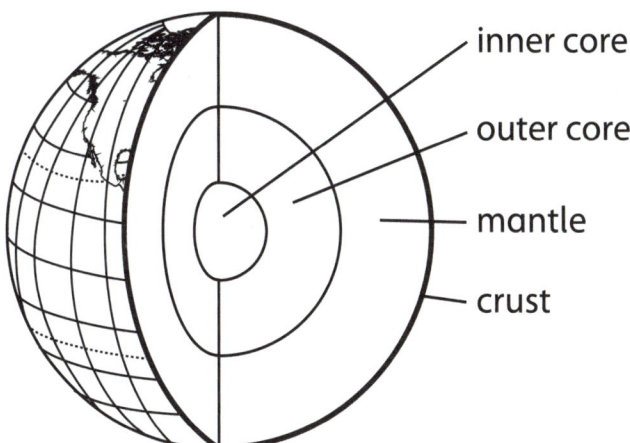

- inner core
- outer core
- mantle
- crust

You will need: different coloured modelling clay, card, string.

1 Use a soft modelling material to make a model of the Earth.
   Choose different colours for the inner core, outer core, mantle and crust.

2 Plan how thick you will make each layer.

3 Make your model. Make sure people can see all of the layers inside.

4 Add labels to your model – use label cards and string.

5 Display your model in the classroom.
   Walk around and see all of the other models.

6 Write down two ways you could improve your model.

Types of rock

Identifying rocks

1 Choose two or three rocks.

2 Work through the questions in the key to help you name each rock.

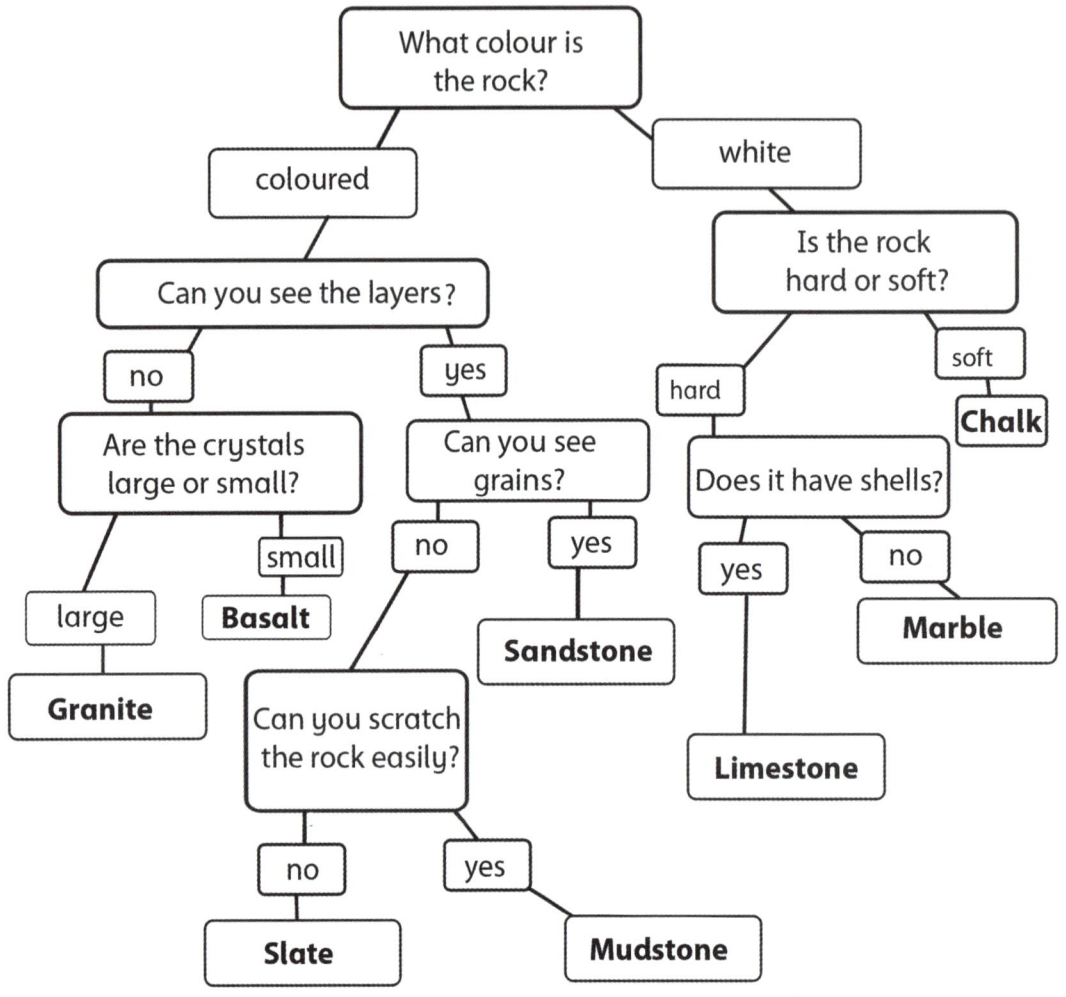

What colour is the rock?

coloured

white

Can you see the layers?

Is the rock hard or soft?

no

yes

hard

soft

Are the crystals large or small?

Can you see grains?

Does it have shells?

Chalk

small

no

yes

yes

no

large

Basalt

Sandstone

Marble

Granite

Can you scratch the rock easily?

Limestone

no

yes

Slate

Mudstone

Hint: A grain is a very small piece of a hard substance. Grains are glued together in some rocks.

3 The rocks I identified are:

Label the rock cycle

1 Look at the diagram of the rock cycle.

2 Complete the labels in the diagram. Use the words in the box. You will need to use one word more than once. Look at page 33 of your Student Book if you need help.

> crystallisation   erosion   igneous   magma   melting
> metamorphic   sedimentary   sedimentation   temperatures

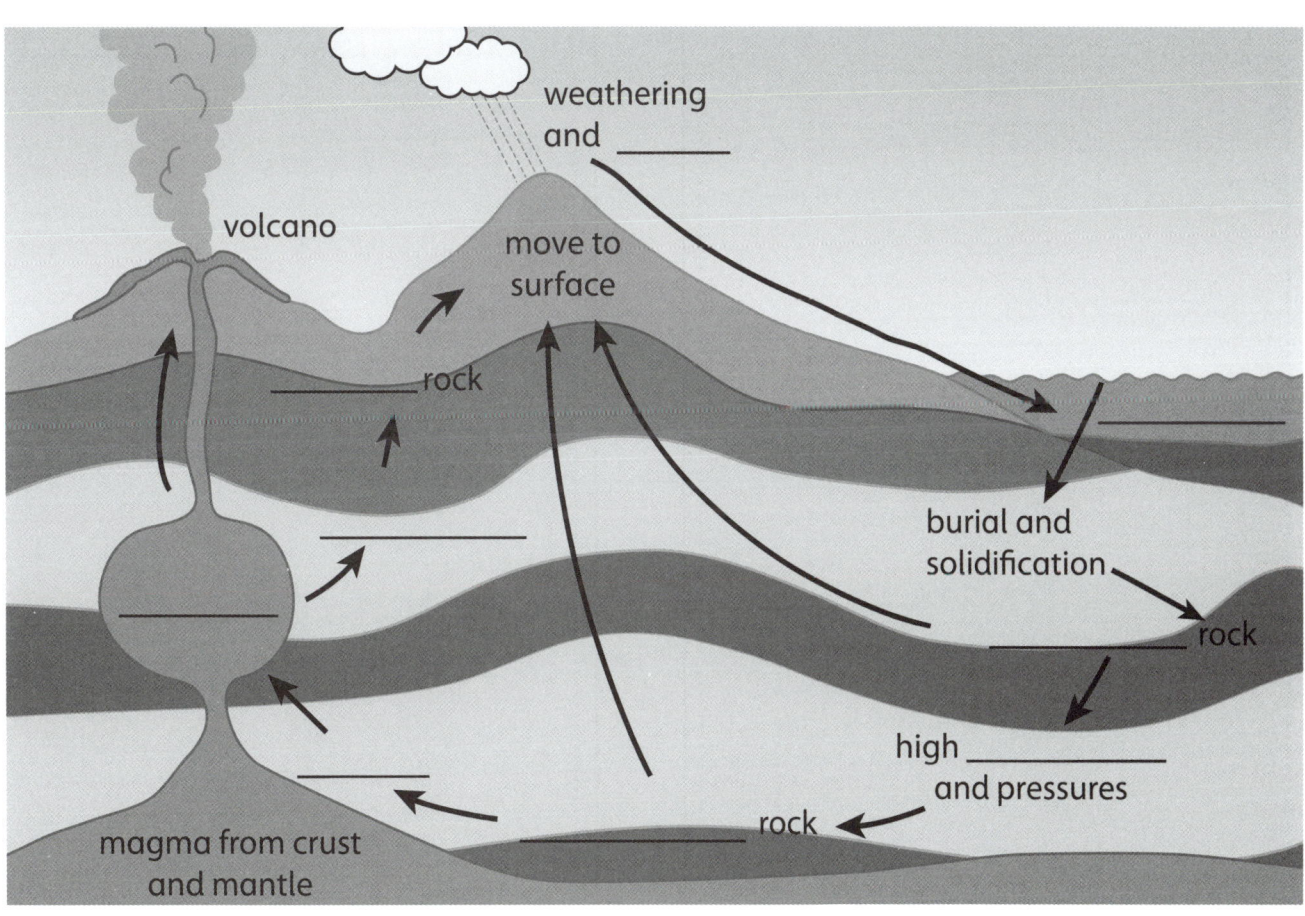

weathering
and _____

volcano

move to
surface

_____ rock

_____

_____

_____

burial and
solidification

_____ rock

high _____
and pressures

_____ rock

magma from crust
and mantle

How fossils form

Fossil field trip

Fossils can be found in a natural history museum or a geology museum.

Visit a museum in your area. Look for different fossils.

1 Walk around the museum and find the area that displays fossils.

2 Choose two different types of fossils.

3 Draw each fossil and then fill in the information cards below.

Name of fossil: _____

How old is the fossil? _____

Where was it found? _____

What type of rock is it found in?

_____

Name of fossil: _____

How old is the fossil? _____

Where was it found? _____

What type of rock is it found in?

_____

Fossil hunt

You are going to investigate an area for fossils.

Your teacher will give you a sand tray. You will search for the hidden fossils.

Your teacher will show you how to make quadrat squares. These will help you record exactly where the fossils are.

1 Make the quadrat squares. Tape string across your tray to make 25 same-sized squares.

2 Give each square an identification number. For example, A1, A2, A3 and B1, B2, B3, as in the grid below.

Warning! Do not let the string touch the sand.

3 Draw a copy of the grid pattern similar to the one below.

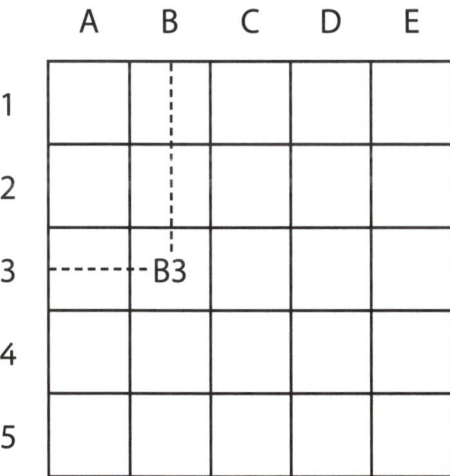

4 Search for the fossils carefully using a small trowel or brush. Each time you find one:

- draw on your grid pattern to record where you found it
- gently remove it and draw a picture of it.

5 Write a report of your fossil hunt. Include:

- what you found
- exactly where you found it.

Rocks as building materials

Building materials

1 Look at the pictures.

 2 Read the words out loud to someone else.

stone wall

concrete path

wooden decking

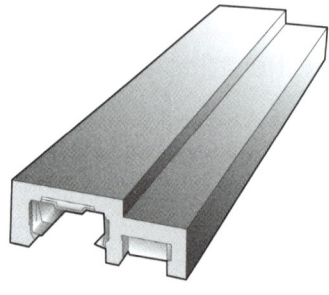

plastic door frame

brick wall

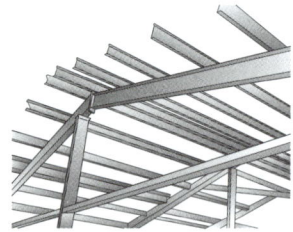

metal girders

3 Complete the table below.

Write the names of all the **human-made** building materials in the correct box.

Write the names of all the **natural** building materials in the correct box.

Human-made building materials	Natural building materials

Building materials survey

Think of all the materials you have learned about.

Which building material is used the most?

Look at the materials used in buildings in your neighbourhood.

1 Predict which material is used the most.

I think _____ is used the most.

2 Use the tally chart below to record your results.

Every time you see a material draw a line in the correct column. This is a tally mark.

Draw your tally marks in groups of five like this:

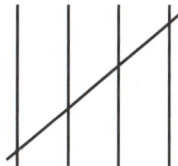

Wood	Brick	Concrete	Metal	Stone	Plastic	Glass

3 Complete the sentences.

My prediction was correct / incorrect.

_____ is used the most in the buildings I saw.

4 Your home is made of different kinds of materials.

Write the ones that are used the most.

My home is made of _____

_____.

More about uses of rocks

Using rocks

Draw a line from the type of rock to how it is used. One has been done for you.

Name and properties　　　　　　　　**How the rock is used**

Sandstone – strong and not easily broken down

statues

Marble – white, attractive and easily shaped and polished

steps, fronts of buildings

Slate – waterproof and breaks easily into sheets

walls of buildings

Granite – very hard, waterproof and attractive

for writing on blackboards

Chalk – white, soft and wears away easily

roof tiles

Testing rock hardness

You will need: three rock samples, a small hammer, a small nail, sand paper.

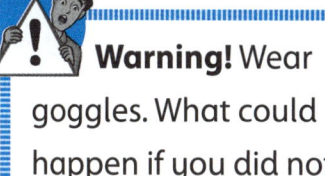

Warning! Wear goggles. What could happen if you did not do this?

You are going to test to see how hard the rocks are.

Test 1

a Gently tap the rocks with a small hammer. See if any grains or pieces fall off.

b Record what you saw:

Broke easily = 1   Difficult to break = 2   Did not break = 3

Test 2

a Try to scratch each rock with a small nail. See how easily the rocks scratch.

b Record what you saw:

Scratched easily = 1   Difficult to scratch = 2   Did not scratch = 3

Test 3

a Rub the rocks with the sand paper. See if the rock is rubbed away easily.

b Record what you saw:

Rubbed away easily = 1   Difficult to rub away = 2   Did not rub away = 3

Stretch zone

What are the most common uses for oil and gas?

_____

_____

What is soil?

What is in soil?

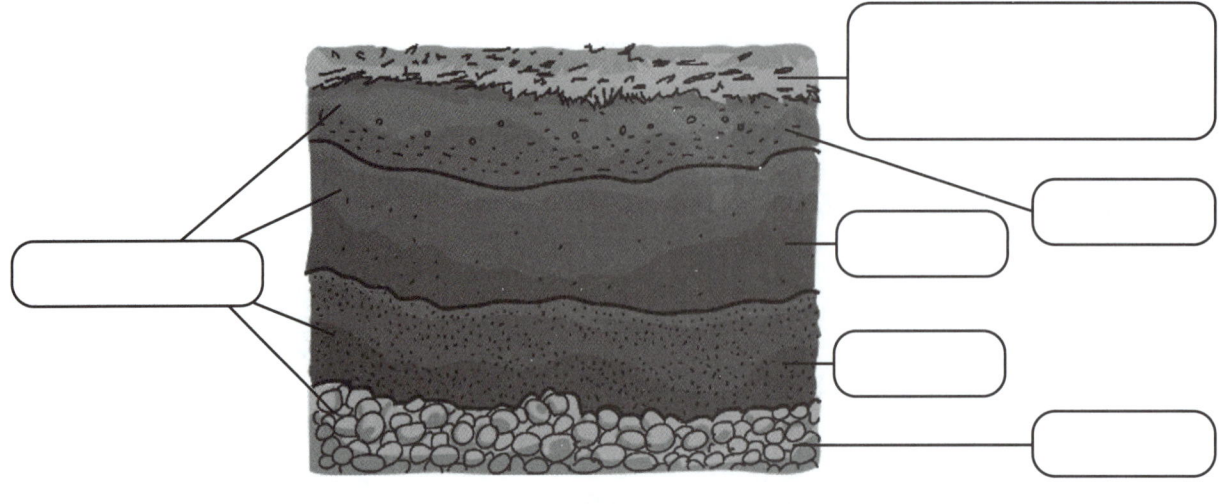

air spaces   clay   dead plants and animal materials
pebbles   sand   water

1 Label the different things found in soil. Use the words in the box.

2 What is the name given to the dead plants and animal materials in soil?

_____

3 Why is soil so important to life on Earth?

Write down two of your ideas.

_____

_____

_____

_____

Investigating different soils

This activity supports the investigation on page 41 of your Student Book.

This activity supports the investigation on page 41 of your Student Book.

You will be given four different soil samples to test.

You will need: four jars, water, a spoon, four soil samples.

1   Set up four jars of water.

2   Add the same volume (amount) of water to each jar.

3   Put five spoonfuls of the first soil sample into one of the jars.

4   Label the jar so you know which soil you have added.

5   Stir the soil and water mixture.

6   Let it settle. This can take up to an hour.

7   Do the same thing with the other three soil samples.

8   Draw what each sample looks like. Use the pictures below to help you.

9   Label your drawings. Use the box to help you.

clay   humus   pebbles   sand

10   How were the soil samples the same?

_____

11   How were the soil samples different?

_____

Different types of soil

1 Read the names of the different types of soil.

2 Then read the descriptions of the different types of soil.

3 Draw a line from each name to the correct description.

chalky soil

- dark brown
- mixture of sand, clay and dead animals and plants
- water drains well
- full of the chemicals needed by plants

clay

- light brown with white pieces
- lots of holes, full of air
- water drains quickly

loam

- a light colour
- lots of holes, full of air
- water drains quickly
- feels dry

sandy soil

- orange or grey
- very few holes, so not much air
- water drains slowly
- feels damp or wet

Soil investigation

Circle the things that seeds need to start growing:

light nitrogen oxygen soil warmth water

Which soil is the best for plants?

You will use two types of soil and two seeds.

Predict which soil will help plants to grow the best.

I predict that the soil labelled _____
will be best for growing the seed.

You will need: two plant pots, two types of soil, two seeds, water.

1 Carefully put a seed into each pot of soil.

2 Cover the seed with soil and water it.

3 Place the pots in a warm and sunny place.

4 Look at your pots every day.

5 Remember to water the soil but do not water it too much.

It might take weeks to complete this investigation. Remember to water the soil and look at the seeds every day.

6 Record what you see.

_____

_____

_____

_____

What I have learned about rocks and soil

What went well

1 Think about what you have learned.

2 Talk to a friend about something that went well in this unit.

3 Tick ✓ the boxes to rate yourself.

I can identify the layers of the Earth's crust.	That's easy. ☐ That's challenging. ☐	Pages 30–31
I can name some types of rock and their uses.	That's easy. ☐ That's challenging. ☐	Pages 32–33
I can describe how rocks and fossils are formed.	That's easy. ☐ That's challenging. ☐	Pages 34–35
I can compare and group rocks using their properties.	That's easy. ☐ That's challenging. ☐	Pages 36–39
I can recognise and name different types of soil.	That's easy. ☐ That's challenging. ☐	Pages 40–43

If you want to know more or need to check, go back to the pages in your Student Book.

Investigate like a scientist

Investigate the uses of rocks

You will investigate if rocks are broken down by acids.

You will use a weak acid such as lemon juice or vinegar.

Week 1:

1   **Weigh three stones you are given. Record their weights in the table below.**

2   **Add each stone to a jar of the weak acid. Leave for a week.**

Week 2:

3   **Wash the stones and dry them.**

4   **Weigh them again. Record their weights in the table below.**

Type of rock (e.g. sandstone, limestone, granite)	Weight at the start (grams)	Weight after a week in acid (grams)

5   **What do the results show you?**

Stretch zone

Rain water can be a weak acid. What does this tell you about why some rocks wear away?

Key words

 Some of the key words for this unit are written below.

There is a problem. Someone has spilt something on the words.

flow r l f poll

ot se

1 Try to work out what each word is.

2 Write each word clearly below. The first letter of each has been done for you.

f __ o __ __ r

l __ __ f

p __ ll __ n

r __ __ t

s __ __ d

Crossword

 Complete the crossword by answering the clues. The words in the box will help you.

compost flower grow leaf
root stem water wilt

Clues:

Across

3   This is where the plant makes food using energy from the Sun.

5   This is how plants get bigger.

6   This can be added to soil to help plants to be healthy.

8   This is needed by plants or they will wilt and die.

Down

1   This is when plants start to fall over.

2   This holds the plant in the soil and takes in water.

4   This attracts insects and makes seeds.

7   This holds the plant upright and helps water travel around the plant.

Make a model of a plant

1 Label the roots, stem, a leaf and a flower on the drawing.

2 Make a model of a plant using a range of materials. When you choose your materials, think about the function (the job) of each part of the plant.

- **Roots:** you could make the roots from string or wool. This material will help the plant take in water from the soil where it is growing.

- **Stem:** you could make the stem from drinking straws. These are like the tiny tubes that transport water and nutrients around the plant.

- **Leaves:** paper towels make good leaves. If you hold a paper towel up to the light, you will see tiny holes. These are like the tiny holes in leaves that allow gases to move into the plant and allow water and gases to leave.

- **Flowers:** you could make the flowers from coloured tissue paper, paper or fabric. Make them bright and colourful to attract insects.

3 Put your plant in a container or pot filled with sand or soil. Make sure your plant is standing up straight and does not fall over.

Flowering plant exhibition

Use the models of the flowering plants that you have made to make a flowering plant exhibition.

1   Show visitors to the exhibition your model of a flowering plant.

- Point out (identify) the flower, roots, stem and leaves.
- Tell the visitors about the function of each part of the flowering plant.
- Explain why you chose the materials for each part. Make sure you link your reasons to the function of each part.

2   Complete the table below to help you prepare for the visitors.

Part of the plant	Material used	Reason for using the material
roots		
stem		
leaves		
flowers		

3   Give each visitor four sticky notes. Ask them to write the name of each part of the plant on a separate sticky note. Ask them to stick the labels onto the correct parts of your model.

Did they get them all correct? Circle the answer.   **yes   no**

4   Write down two ideas to make your model better.

_____

_____

Flowering plant wordsearch

1 Find and circle the four main parts of a flowering plant in the wordsearch.

Hint: Not all of the words are written up and down or across!

t	u	r	x	k	m	l
f	s	k	o	f	g	h
c	t	y	c	o	j	s
d	e	a	s	v	t	l
a	m	d	l	o	y	l
l	j	u	b	p	x	e
f	l	o	w	e	r	a
l	z	w	r	h	p	f

2 Write the function of each part of the plant.

3 Write down what the parts would look like in an unhealthy plant.

Plant rescue

A person has been trying to grow some pot plants.

Some have grown into healthy plants.

Some are unhealthy.

1 Draw a circle around any of the pictures that show healthy plants.

2 Draw a star (☆) next to any unhealthy plants.

3 Write down the clues you used to decide which were unhealthy.

4 Write a short message to the person to tell them how each unhealthy plant could be made healthier.

 Reflect on what you have learned about unhealthy plants.

Do plants need water?

Do plants need water to grow?

 This activity supports the investigation on page 52 of your Student Book.

You will investigate whether plants need water to grow.

What you will investigate

Your question is: 'Do plants need water to grow?'

1 What do you predict will happen to the plants in your investigation?

_____

2 Why do you think this?

_____

_____

_____

Planning the investigation

3 I will need: *soil*, _____.

4 I am going to change _____.

5 I am going to keep _____ the same.

6 I am going to measure _____.

Making observations

7 What will you use to measure with? _____

Measuring plants

This activity supports the investigation on page 53 of your Student Book.

There are different ways to measure how well plants are growing.

Try these two methods in your investigation.

1 **Keep the plant in the soil.** Measure the height of the plant from the soil to the highest tip. Use a ruler.

2 **Remove the plant from the soil.** Wash the roots. Measure from the tip of the longest root to the highest part of the plant.

3 What is the main advantage of the second method?

4 What is the main disadvantage of the second method?

Do plants need light?

Plants and light investigation

 This activity supports the investigation on page 54 of your Student Book.

You are going to observe seedlings growing in the light and in the dark.

1 Record your observations of your seedlings that are growing in the light and in the dark in the table below.

Day	Description of seedlings grown in the light	Height of seedlings grown in the light (cm)	Description of seedlings grown in the dark	Height of seedlings grown in the dark (cm)
1				
2				
3				
4				
5				

2 Draw a graph to show the heights of the seedlings grown in the light and in the dark.

 Think about what you have learned about light, dark and growth.

1 Make as many words as you can using the letters from the word below:

photosynthesis

2 Write the words here. One has been done for you.

is

3 Count the number of words you made.

4 Find two more words that begin with 'photo-'. You can use a dictionary for this too.

Write the words here.

The importance of roots

Grow a bean seed

 You are going to grow a bean seed in a jar.

1 Set up the growing jar as shown in the diagram.

You will need: a glass jar, paper towels or blotting paper, an uncooked kidney bean or other bean, water.

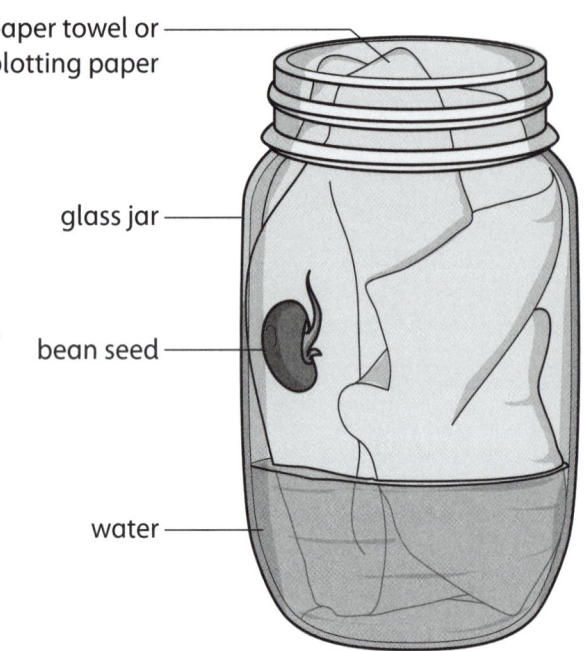

paper towel or blotting paper

glass jar

bean seed

water

2 Place the jar on a windowsill where it will get lots of sunlight.

3 Observe your bean seed every day for a few weeks.

4 Draw your observations on a piece of paper every few days. Label the roots.

 **Stretch zone**

Why do the roots grow downwards?

_____

Why does the stem grow upwards?

_____

 Explain your ideas to someone else.

How water moves through plants

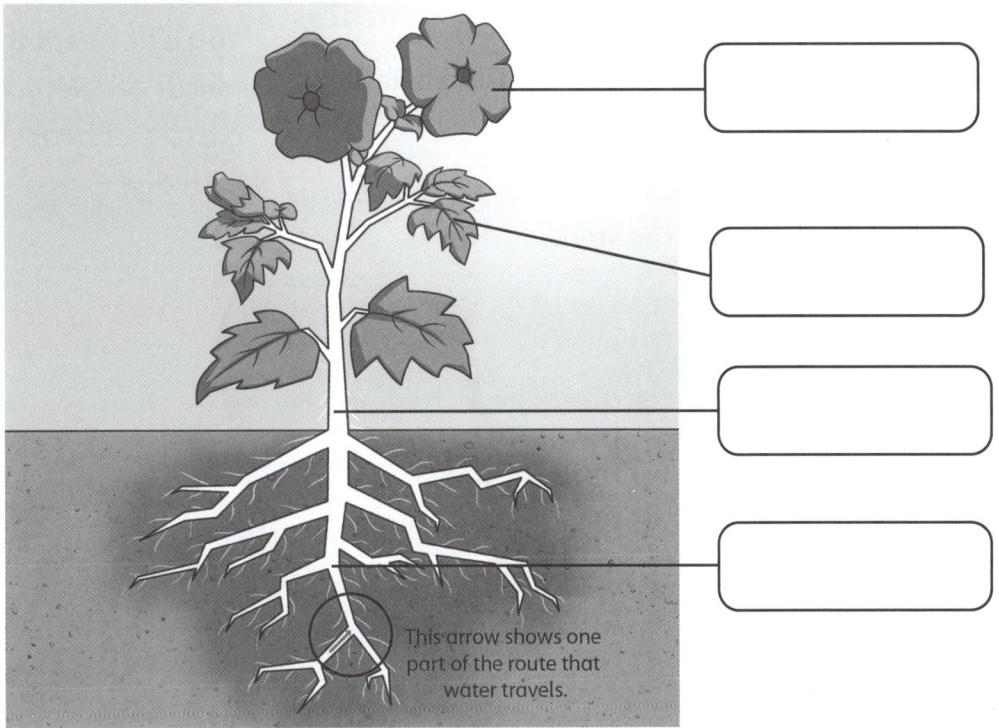

This arrow shows one part of the route that water travels.

1   On the diagram, draw arrows to show the route that water travels through the plant.

    One arrow has been drawn for you.

2   Label the different parts of the plant.

3   Roots are important to plants.

    Read the list below. Tick ✓ the two things that roots do.

Keep the plant secure in the ground ☐

Make food ☐

Get water from the soil ☐

Give the plant its colour ☐

Get energy from the Sun ☐

The importance of stems

Do leaves affect how a plant transports water?

 Does the number of leaves affect how water is transported?

You have already carried out an investigation using celery (see page 58 of your Student Book).

Now you will investigate if the number of leaves on the celery affects how water is transported in the stem.

You will need: three pieces of celery, three glasses or beakers, water with dye.

1   Take one piece of celery and follow the steps in the investigation on page 58 of your Student Book.

2   Take another piece of celery and pull off half the leaves. Then follow the same steps.

3   Take another piece of celery and pull off all the leaves. Then follow the same steps.

4   Predict what you think will happen.

_____

_____

5   Check what happens to all the celery over the next two days. Use a ruler to measure how far up the stem the dye has moved.

6   On a separate piece of paper, draw a table to record your observations.

Does the number of leaves affect the way the dye moves through the stem? Was your prediction correct?

_____

Stems for support

Stems allow water to move from the roots to the leaves.

They also offer support to the plant. This keeps leaves and flowers above the ground.

1 Set up a model stem as shown in the picture.

   You can use plastic straws or dry spaghetti to act as the stem.

2 Stick the model stem into a pot of sand and add four clay leaves.

   Gently tip the pot and write down what happens to the stem.

3 Add two more leaves along the stem and a model flower to the end of the stem.

   Tip the pot again. What happens?

4 Make another model stem but this time stick together six straws or strands of spaghetti.

   Add six leaves and one flower.

   Tip the pot. What happens?

5 Now add four more leaves.

   Tip the pot. What happens?

Plant parts work together

The power of roots

Roots can be very strong. They can even force their way through bricks or road surfaces.

1 Look for strong roots. Look around the area where you live.

Try to find some roots that are growing like the example in the photograph.

2 Have the roots caused any damage?

Draw the roots you find. Show any damage they have done.

Plant expert's report

1   Complete the plant expert's report on what is wrong with the plant in the picture.

You can use words from the box below to help you. One has been done for you.

The leaves are dead so they cannot <u>photosynthesise</u>.

The stem is wilted so it cannot _____ the plant.

The roots are _____ _____ so it is difficult for the plant to take in _____.

~~photosynthesise~~   pot bound   support   water

2   Look at these flowering plants and plant pots. Match each plant with the best sized plant pot. Colour in the plants so they look healthy.

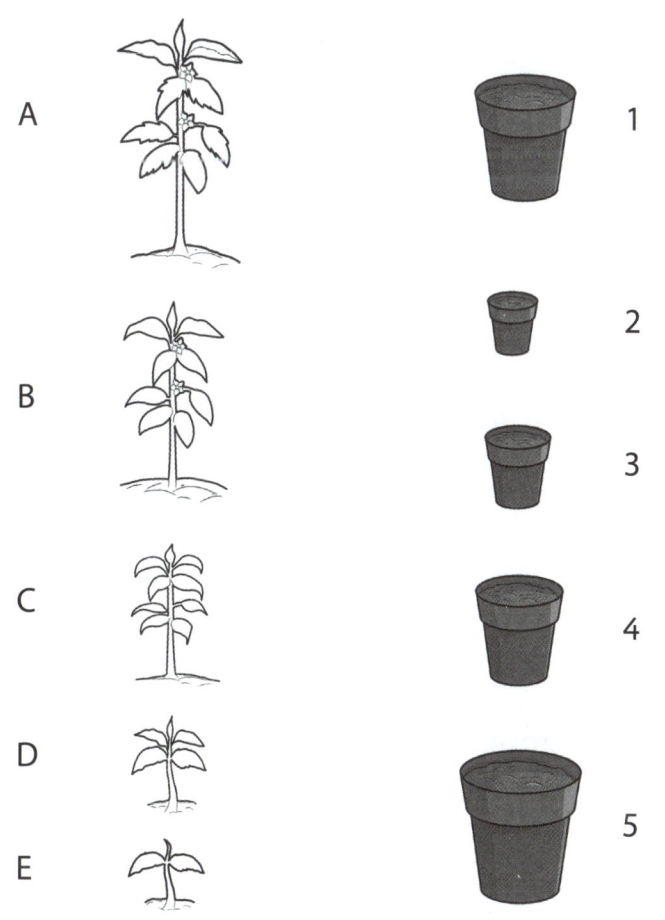

A

B

C

D

E

1

2

3

4

5

Tree survey

You will visit an outside area.

1 Carry out a survey of all the trees you see.

Use books or the internet to help you identify the trees and what they can be used for.

Record the number of trees of each type that you find.

2 Find the largest tree of each type. Measure the size of the trunk. Look at the pictures to show you how to do this.

3 Use the table below to record your results neatly.

Name of tree	Circumference of tree trunk (cm)	Number of trees	What this tree is used for
Willow			To make baskets

4 Which tree had the largest diameter? _____

5 Which trees were growing well? Did they have a lot of space?

6 Were any trees crowded together? Were they growing well?

Unfair competition

This activity supports the investigation on page 63 of your Student Book.

1 Predict how well the small plant will grow.

2 What is your reason for predicting this?

3 Observe what happens to both plants for a few days.

Record your observations in the table below.

Day	What does the large plant look like?	What does the small plant look like?
1	Healthy, with firm green leaves	
2		
3		
4		
5		

4 How well did the small plant compete for space, water and sunlight?

Not too hot and not too cold!

Seedling investigation

 In what conditions do plants grow best?

You are going to investigate which combination of light and temperature helps plants to grow best.

1 Half fill each pot with compost.

2 Carefully push a seed into each pot, until the seed is covered with the compost.

3 Water the pots. Place one pot in each of the places listed in the table below.

4 Check the pots every day. A plant should grow in each one.

Predict which plant you think will be the healthiest.

_____

5 Measure how tall each plant grows during two weeks. Count the number of leaves. Observe whether the plant looks healthy or unhealthy. Use the table to record your results.

Place	Height (cm)	Number of leaves	Observations
a cool, dark place			
b cool, light place			
c warm, dark place			
d warm, light place			
e hot, dark place			
f hot, light place			

6 I found that plants grow best in _____.

Make your own greenhouse

Design and make a greenhouse for a plant.

You will need different materials to make a greenhouse:

- cling film and polythene are good materials for the top
- plastic food boxes or trays are good for the container.

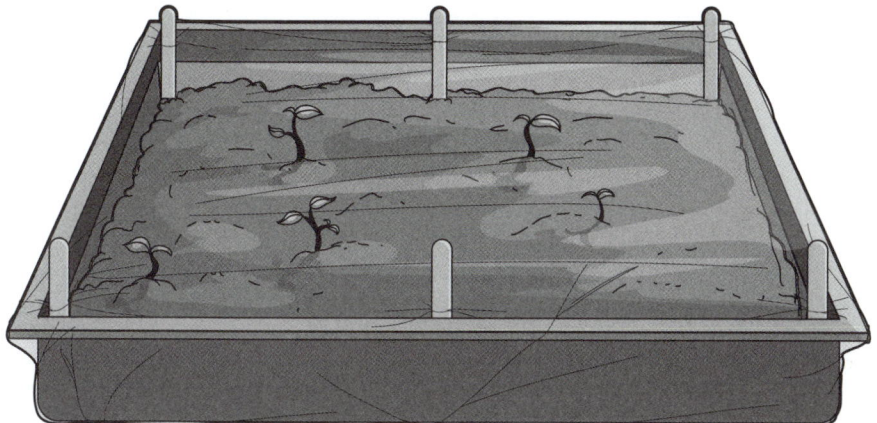

Think about these points.

- What container will you use for the plant to grow in? It will have to keep the soil and water in.
- How big does the container need to be? If you are growing seedlings, the tray does not have to be very deep.
- Does the greenhouse need to let light through? If it does, what material will you use for the top?
- Do you need to remove the top? You might need to check the height of your plant or water it.

Design your greenhouse and then build it.

The parts of a flower

Label the parts of the flower. Use the words from the box.

anther filament ovary petal receptacle
sepal stigma style

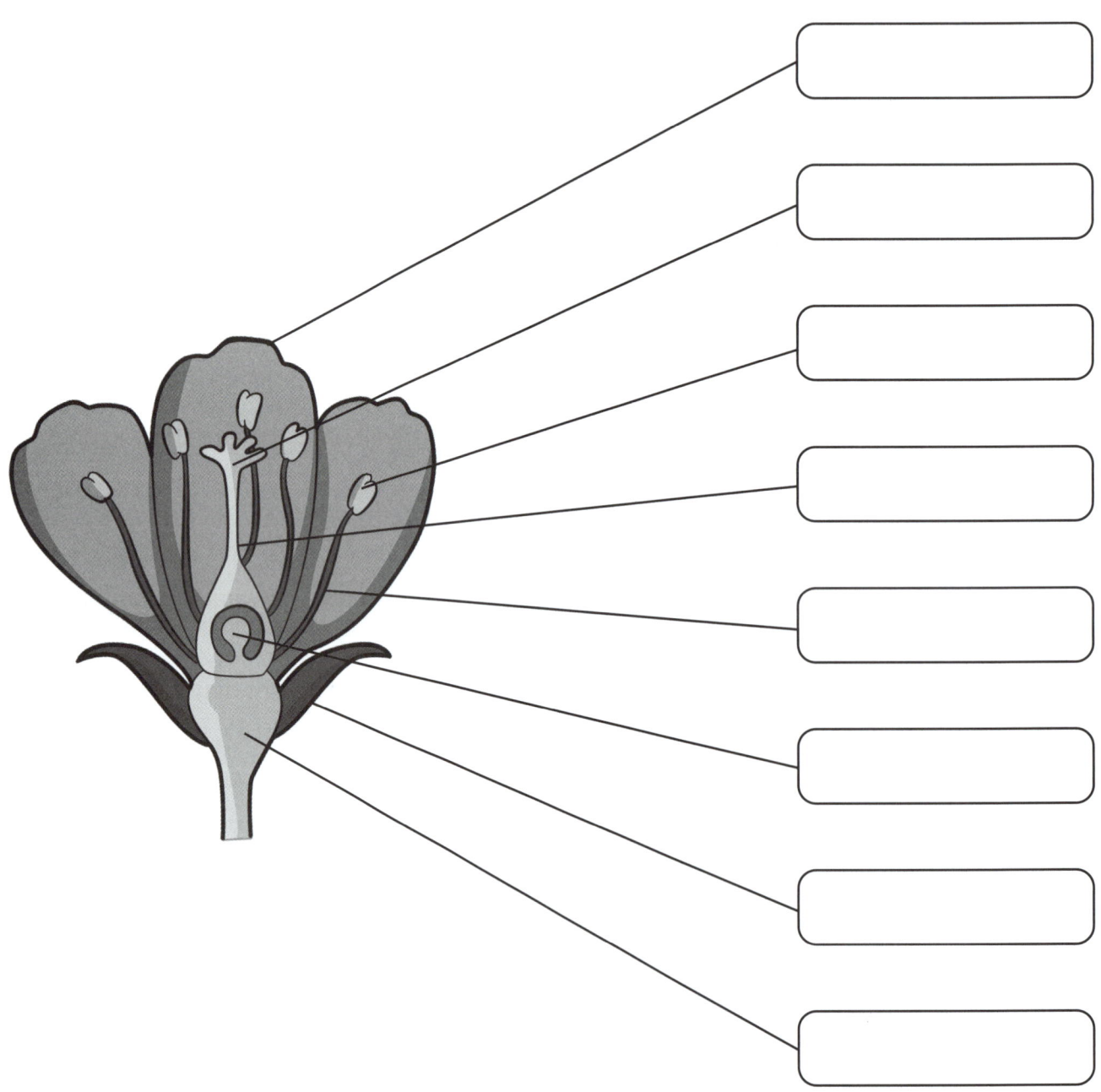

Identify the parts of the flower

This activity supports the investigation on page 67 of your Student Book.

1 Fill in the table below.

Part of the flower	Function – what it does
ovary	
	collects pollen
petals	
anther	
	holds the anther up
	protects the flower before it opens
style	
	thick part of the stem

2 Draw a stamen and a pistil in the boxes below.

Pollination and seeds

Fertilisation

1 Label the diagram. Use the words from the box.

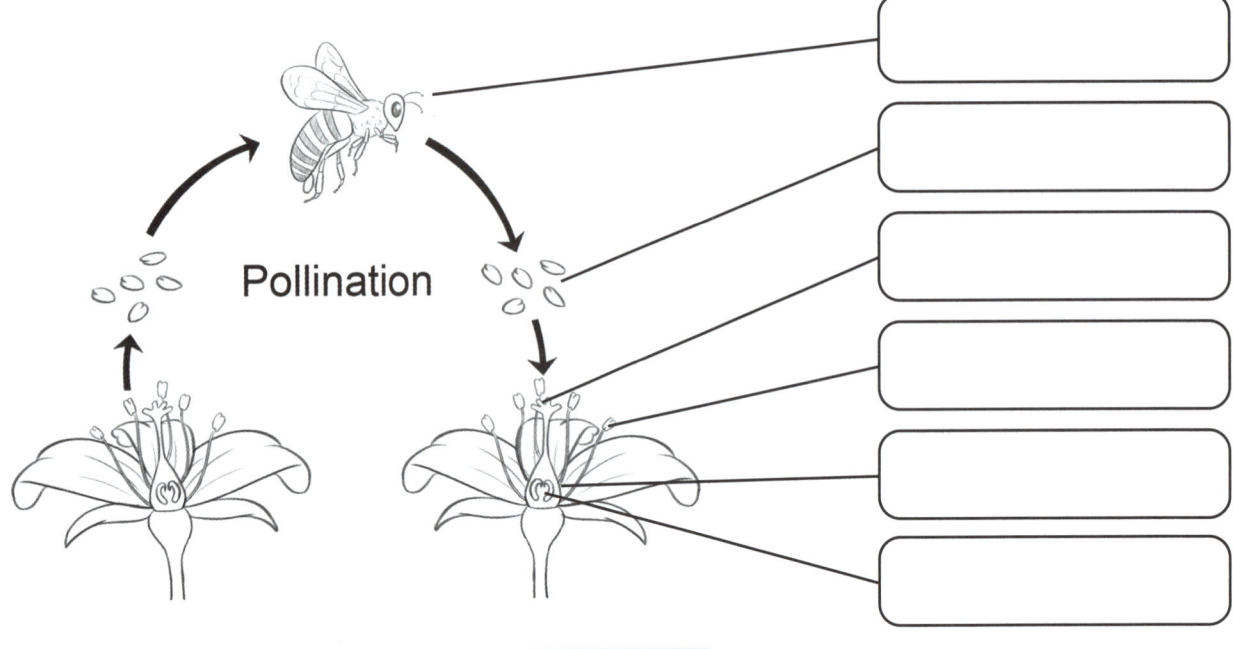

Pollination

ovary ovule pistil pollen pollinator stamen

2 Which part of a flower is shown in the diagram below?

_____

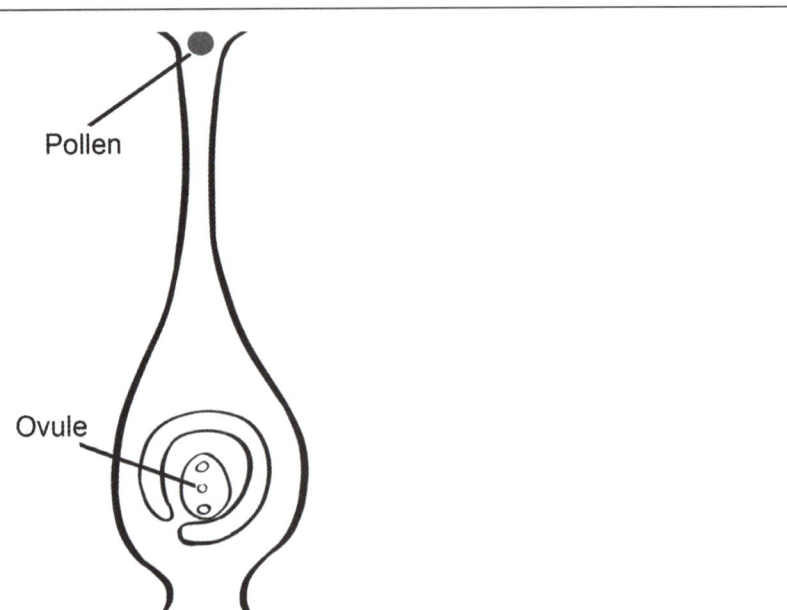

Pollen

Ovule

3 On the diagram above, draw what happens to allow the pollen to fertilise the ovule.

Investigating seed dispersal

1 Which seeds in the pictures above do you think are the most likely to grow and develop into healthy plants? Make a prediction.

2 Plan and carry out an investigation to test your prediction.

   • Use two different plant pots containing compost to sow your seeds.

   • Decide where to place your plant pots.

3 Observe what happens over the next two weeks.

Discuss what you have found out.

What I have learned about flowering plants

 What went well

1 Think about what you have learned.

2 Talk to a friend about something that went well in this unit.

3 Tick ✓ the boxes to rate yourself.

I know that plants have roots, leaves, stems and flowers.	That's easy. ☐ That's challenging. ☐	Pages 48–49
I can explain observations that plants need water and light to grow.	That's easy. ☐ That's challenging. ☐	Pages 50–55
I know that water is taken in through the roots and transported through the stem.	That's easy. ☐ That's challenging. ☐	Pages 56–59
I know that plants need healthy roots, leaves and stems to grow well.	That's easy. ☐ That's challenging. ☐	Pages 60–63
I know that plant growth is affected by temperature.	That's easy. ☐ That's challenging. ☐	Pages 64–65
I know the life cycle of flowering plants.	That's easy. ☐ That's challenging. ☐	Pages 66–69

 If you want to know more or need to check, go back to the pages in your Student Book.

Investigate like a scientist

Making your own plant country

You will make a country made of plants.

1 **Cover the bottom of a seed tray with 5 cm of compost.**

2 **Shape the compost to make hills and valleys.**

3 **Carefully cut the tops off some root vegetables. You could use carrots, swede, turnips or parsnips.**

4 **Plant the tops of the vegetables into the compost. Make sure the cut end is in the compost.**

5 **Add leaves, twigs, shells or stones to make your country look more realistic.**

6 **Place your country in a warm and sunny place.**

7 **Keep the compost damp. How else can you care for the plants?**

_____

8 **Photograph your plant country before and after your plants start to grow.**

9 **What did your vegetables need to help them grow healthily? Make a list.**

_____

_____

_____

_____

_____

_____

_____

_____

Key words

We cannot see forces.

We can see how forces act on objects.

Push and **pull** are examples of forces.

Draw a picture to show an example of each of the forces written below

When I have used a push force

When I have felt a push force

When I have used a pull force

When I have felt a pull force

Introduction

Magnets

1 Find the key words about magnets in the wordsearch. Circle each one in the grid below.

attract compass magnetic
north poles repel south

r	u	p	n	o	r	t	h	t	n
i	o	w	p	a	v	l	m	w	i
k	s	o	u	t	h	u	a	c	c
z	z	s	m	t	x	i	g	o	k
r	j	o	u	r	r	b	n	m	e
e	c	z	f	a	g	u	e	p	l
p	n	y	i	c	g	w	t	a	d
e	l	v	s	t	e	d	i	s	x
l	c	p	o	l	e	s	c	s	m
s	l	u	i	r	o	n	l	g	i

2 Find the two hidden magnetic metals in the wordsearch.

_____   _____

3 Write down one thing you already know about magnets.

Pushes and pulls

Find the forces

1 Look at what is happening in the picture.

2 Label all the pushes and pulls you can see.

pull   push

3 Draw an arrow to show the direction of each force.

4 Discuss your ideas with someone else. You might not agree on them all.

Showing the direction of a force

We can draw arrows to show the direction of a force.

1

This is a drawing of two small boats and a crane. Draw an arrow to show the direction of the force acting on the crane.

2

This is a drawing of a child and a pushchair. Draw an arrow to show the direction of the force acting on the pushchair.

 Stretch zone

Write a label for each force.

Measuring pushes and pulls

Make a forcemeter

 You are going to make and use a forcemeter.

1 Use an elastic band to make your own forcemeter.

You will attach different objects to the elastic band.

This allows you to measure the force needed to lift each object. The longer the elastic band stretches, the bigger (greater) the force.

A forcemeter has a scale to help us measure the force accurately. You can use a ruler to measure the length of the elastic band.

2 Make a scale for your forcemeter. You can use these terms:

- 'large force' – when the elastic band stretches a lot
- 'small force' – when the elastic band does not stretch very much
- 'medium force' – in the middle between large force and small force.

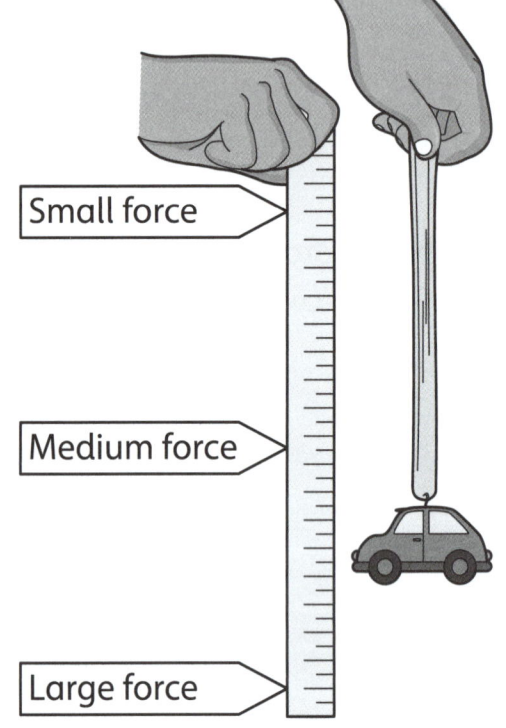

Small force

Medium force

Large force

3 Use your forcemeter to lift small objects in your home.

4 Look at the scale. Observe the size of the force needed to lift each object.

Present your results in the table below.

Object	Size of force (large, medium, small)

Using a forcemeter

This activity supports the investigation on page 77 of your Student Book.

You are going to investigate how force is needed to open doors.

Warning! When you use a forcemeter to pull or push objects, the needle moves. Sometimes it doesn't stop. You have to decide where the needle stays for most of the time and read that measurement from the scale.

1 Look at a forcemeter closely. Practise reading the scale.

2 Start with the door slightly open. Record the measurement in the table below.

3 Record the results for any other doors that you investigate in the table.

Location of the door	Force (N)
room door	
cupboard door	

4 Describe any patterns in your results.

_____

_____

5 What did you find out? Do you need more force to open bigger doors than smaller ones?

_____

_____

Making shapes with forces

Make your own play dough

 You have learned how forces can change the shape of objects. Now you can investigate this by making and using your own play dough.

You will need:

- 2 cups plain flour (all purpose)
- ½ cup salt
- 2 tablespoons cream of tartar
- 2 tablespoons vegetable oil
- 1.5 cups boiling water
- food colouring (optional)
- a few drops of glycerine (optional: it adds more shine!)

1 Mix the flour, salt, cream of tartar and oil in a large mixing bowl.

2 Add the food colouring (optional) to the boiling water. Pour the coloured water into the dry ingredients a little at a time.

3 Stir continuously until you have a sticky, combined dough.

4 Add the glycerine (optional). Then allow the dough to cool down.

Warning! Ask an adult to help you make the dough. The adult will add the boiling water and stir the dough. Do not do this yourself. Discuss why this is important.

Optional means you can choose whether or not to do this.

5 Take the dough out of the bowl and knead it vigorously for a few minutes until all of the stickiness has gone. This is the most important part of the process, so continue until the dough is the perfect consistency.

6 Squeeze, squash, stretch and roll the dough to change its shape.

7 Think about how each type of force changes the shape of the dough.

Using forces to change the shape of modelling clay

This activity supports the investigation on page 79 of your Student Book.

You are going to use forces to change the shape of modelling clay.

1 Use the modelling clay to make a pot.

2 Squash, stretch and squeeze the clay to make it into the shape that you want.

Use the pot to help you to carry out the investigation below.

3 **Planning my investigation: Using forces to change the shape of modelling clay**

I will need _____

_____.

What am I going to change?

_____

_____

What am I going to keep the same (to make this a fair test)?

_____

_____

What am I going to measure?

_____

_____

_____

What am I going to do?

_____

_____

_____

Forces can stop or start things moving

Bouncing ball investigation

 You are going to use forces to control a ball.

You will need: a bouncing ball (a soft foam ball is best for this investigation), a stop watch or a clock or watch with a second hand (to measure the time in seconds).

1 Find a safe place to throw a ball at a wall.

2 Throw the ball at the wall. Is this a push or a pull?

_____

3 Can you throw the ball at the wall so that it bounces back?

4 Try to control how fast the ball bounces back.

- Throw the ball and ask your helper to start the clock or timer.
- Can you make the ball take 5, 4, 3, 2, 1 seconds to return to you?
- Make sure you always stand the same distance away from the wall.

5 Can you find a pattern in your results?

6 How do you change the way you throw the ball to make it come back more quickly?

_____

_____

7 How do you change the way you throw the ball to make it come back more slowly?

_____

_____

 Think about what you have learned in this investigation.

Measuring distance

This activity supports the investigation on page 81 of your Student Book.

You are going to investigate how forces make a toy car change speed and stop.

You need to measure the distance that a toy car travels across the floor.

1 What piece of equipment will you use to measure the distance?

Which one of these is the best to use? Circle your answer.

30-centimetre ruler   **tape measure**   **metre stick**   **steel ruler**

2 Look at the picture of the ruler.

- It is important to start measuring from the beginning of the scale.

- Draw an arrow to show the beginning of the scale on this ruler.

Warning! When you measure the distance the toy car has travelled, make sure you measure to the same point on the car every time. Choose the front or the back of the car and measure to that point every time. This helps to make it a fair test.

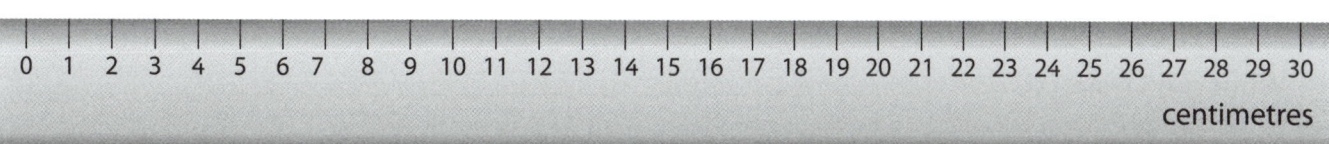

3 Practise measuring using a ruler. Use your ruler to draw lines of these lengths in the space below.

2 centimetres

5 centimetres

12 centimetres

Forces on different surfaces

Vehicles travel further on some surfaces

This activity supports the investigation on page 82 of your Student Book.

You are going to investigate how different surfaces affect the distance that a toy car travels.

What do you predict will happen to the distance the car travels on different surfaces?

Predict what will happen

I think that _____

_____.

My reason is that _____

_____.

My investigation plan

I will need: _____.

What am I going to change? _____

What am I going to keep the same? _____

What am I going to measure? _____

What am I going to do? _____

_____

I will be careful of _____.

My drawing of what I will set up

Surface investigation

Look at the picture. How does the ice skater move so quickly?

You are going to investigate how easily objects move on a surface.

1 Place some different small objects on a work surface or table.

2 Push and pull the objects to move them across the surface. Is it easy to move them?

 • How hard do you have to push and pull the objects to make them move?

 • You could use the forcemeter you made to measure the force needed to move each object.

3 Now use an ice cube. Carefully place each object on the top of the ice cube so that it is balanced.

4 Push and pull the objects while they are on top of the ice cube to move them, as you did before. Do the objects move more easily now?

Draw a table to record your results.

5 Compare the movement of the objects with and without the ice cube. How does the ice affect friction on the surface?

Friction

Which shoe to use?

Look at the pictures of footwear. Look at the list of uses.

Which footwear is best for each use?

1 Draw a line to link each piece of footwear to its use.

ice skating

dancing

walking

ice climbing

playing football

running on a track

running on mud

2 How do each of the shoes change the amount of grip?

_____

_____

Which shoe has the best grip?

You are going to investigate which shoe has the best grip.

1 Measure 1 metre across a flat space.

You can mark this using a piece of tape at the start and another piece at the end.

2 You will test some different shoes. Look at each one. How much force do you think each shoe will need to move? The more force that is needed the better the grip of the shoe.

The shoe that will need the most force is the:

_____.

The shoe that will need the least force is the:

_____.

3 Place the shoe at the starting point. Attach the forcemeter to the shoe.

4 Pull the shoe gently across the surface to the end point. Measure the force on the scale of the forcemeter while you are pulling the shoe.

5 Record this in a table of results.

Type of shoe	Force needed to move the shoe (N)

6 a Did your results agree with your predictions?

_____

_____

b Were there any surprises in your results?

_____

_____

c Did some shoes use more force than you expected?

_____

_____

Forces can change the direction of moving objects

Rolling a ball

 You are going to explore how a ball moves. You will need to sit in front of a wall.

You will need: two tennis balls or soft foam balls, a tape measure or a ruler.

1 Practise rolling the ball so it hits the wall and bounces away.

2 Measure how far it bounces away.

3 Who can get the ball closest to the wall?

Roll the ball so it stops as close to the wall as possible. Have three turns each. Measure the distance from the ball to the wall and record your results.

4 Who can make the ball stop 1 metre away from the wall?

- Roll the ball so it bounces off the wall and stops as close to 1 metre away from the wall as possible. Have three turns each.

- Measure the distance and record your results.

- Draw a table to record your results.

 Think about what you had to change to get the ball as close as possible to the point you were aiming for.

Forces and moving objects

You are going to think about how objects move.

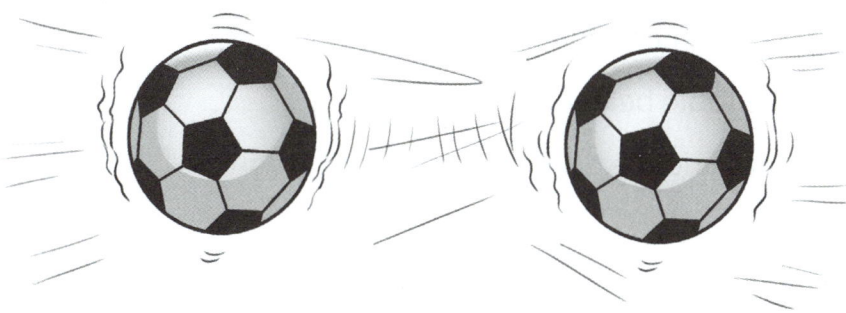

1 What do you think is happening when an object hits another object?

2 What forces are acting on the objects?

3 Complete the following sentences using the words in the box.

When two moving objects hit _____ _____ the opposite forces can make the objects change _____.

When two objects hit each other _____ slows the objects _____.

direction down each other friction

4 Read out the words in the box. Discuss what each word means.

Is it magnetic?

Which objects are magnetic?

Which objects are magnetic?

You will need: a magnet.

1   Look around the room. Choose six objects that you predict will be attracted towards a magnet (are magnetic).

Write the names of the objects in the table below.

2   Choose six objects that you predict will not be attracted towards a magnet (are non-magnetic).

Write the names of the objects in the table.

The six objects I predict are magnetic	The six objects I predict are non-magnetic

3   Use a magnet to test your predictions.

How many of your predictions were correct?
Tick ✓ all the objects you predicted correctly.

4   Magnets can be very useful at home. Write three examples.

1 _____

2 _____

3 _____

How can we identify the poles of magnets?

 You are going to see if you can identify the poles of magnets when they are covered up.

Ask an adult to set up some magnets so you cannot see the poles. Any of the magnets could be arranged with the North-seeking pole towards you or away from you.

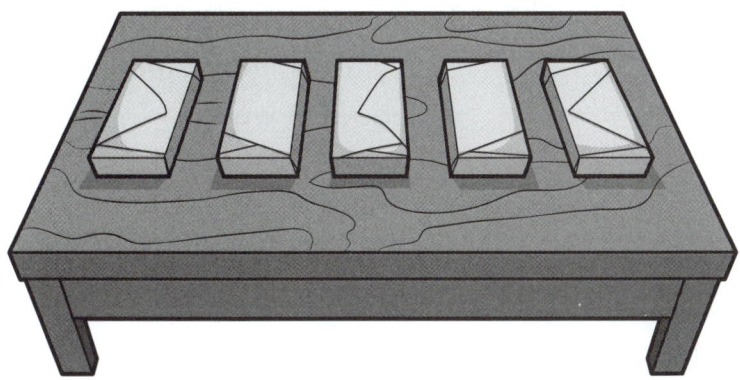

1 Look at an uncovered magnet.

   How can you find out which pole is which on the covered magnets without taking the paper off?

2 Test each magnet in turn. Record your findings by writing N or S on the paper covering at the correct end.

3 Ask an adult to unwrap the magnets so you can check your answers.

   How many of the five did you get correct? _____

4 Which law of magnetism did you use to help you work out your answers?

   _____

Using magnets

Make a compass

 Magnets have many uses. The needle of a compass is a small magnet with a low magnetic strength. You are going to make your own compass.

You will need: a magnet, a sewing needle or nail, a compass, a black marker pen, a plastic bottle top, glue or sticky tape, a bowl of water.

1 Make the compass needle by stroking the magnet along a sewing needle or nail.

Warning! Take care with the needle. It has a sharp point. What could happen if you were not careful?

2 Always stroke the needle in the same direction – not backwards and forwards. Stroke it about 12 times.

3 Mark the pointed end of the needle using a black marker pen.

4 Fix the needle to a plastic bottle top using glue or tape.

5 Float the bottle top and needle in a bowl of water.

6 When the needle stops spinning, notice which direction the needle points. This will be North–South.

7 Use a compass to find out if the black end of your needle is pointing North or South.

Once you know which directions are North and South, you can work out East and West.

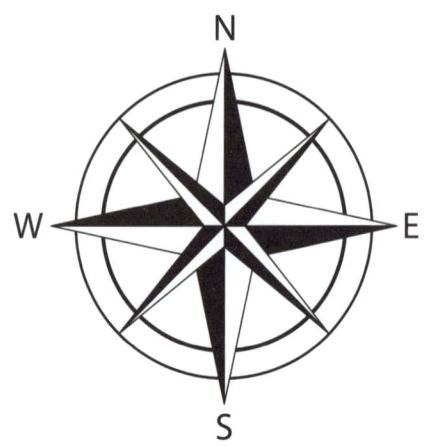

8 Use your compass to find:

- a large or important building to the West of your room. This building is _____.

- a large or important building to the North of your room. This building is _____.

All about magnets

1 Fill in the missing words. One has been done for you.

Some objects __attract__ other objects to them. They are called _____.

A _____ has two ends. One end is the _____-seeking pole and the other end is the _____-seeking pole. The law of magnetism is: like poles _____ and opposite poles attract.

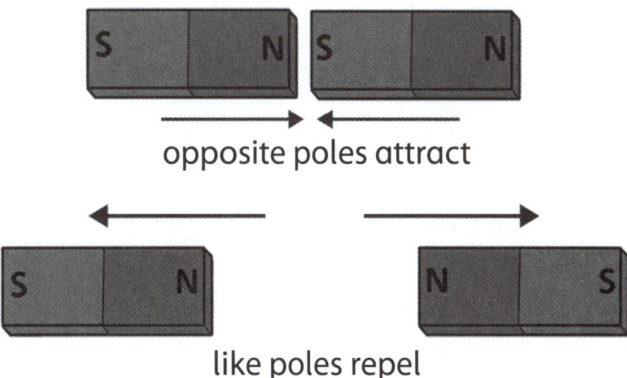

opposite poles attract

like poles repel

Iron, _____ and cobalt are magnetic materials. All other _____ and non-metals are non-magnetic. These materials are not _____ to a magnet. There are many uses of magnets.

> ~~attract~~ attracted magnet magnets metals
>
> nickel North repel South

2 How many different kinds of magnets have you learned about? Draw a diagram of all of the magnets you know in the box below.

Magnets have poles

Buried treasure

 Use a magnet to find treasure!

You will use a container full of sand. Hidden in the sand is buried treasure.

1 Search for the treasure by moving a magnet over the sand close to the surface.

2 Use the magnet to collect objects. Place all the objects you collect into another container.

3 Continue until your magnet cannot find any more objects.

4 Then use your hands to find the rest of the objects in the sand. Place all these objects in another container.

 Why did the magnet not pick up these objects?

5 Fill in the table below by writing the objects you found with the magnet and the objects you did not find with the magnet.

Objects found with the magnet	Objects not found with the magnet

Design a recycling plant

Magnets are used in many different places. Recycling plants use them to separate magnetic materials from non-magnetic materials.

You are going to design a device to sort waste. You will use a magnet in your design.

Your device must be able to separate magnetic waste, such as cans, from non-magnetic waste, such as paper and bottles.

Think about these questions.

- Where will you position the magnet?
- How will the waste move to the magnet?
- How big will the magnet need to be?
- Where will the separated waste go next?

1 Design your device in the space below.

2 Write labels and notes to explain how your device works.

Investigating the poles of a magnet

Why does the Earth have a North Pole and a South Pole?

 1 Fill in the gaps in the description. Use the words in the box.

At the _____ of the Earth there is a liquid core. The _____ is made of molten

metals that are _____. It creates a massive magnetic _____ force at each end.

This force wraps around the Earth's surface and creates the Earth's magnetic _____

with a North end and a _____ end. The North end of a bar magnet is attracted to the

_____ Pole. The South end of a bar magnet is attracted to the South _____.

> centre core field magnetic North
>
> Pole pulling South

2 Look at the diagram. Fill in as many of the labels as you can. If you need help, look at page 93 of your Student Book.

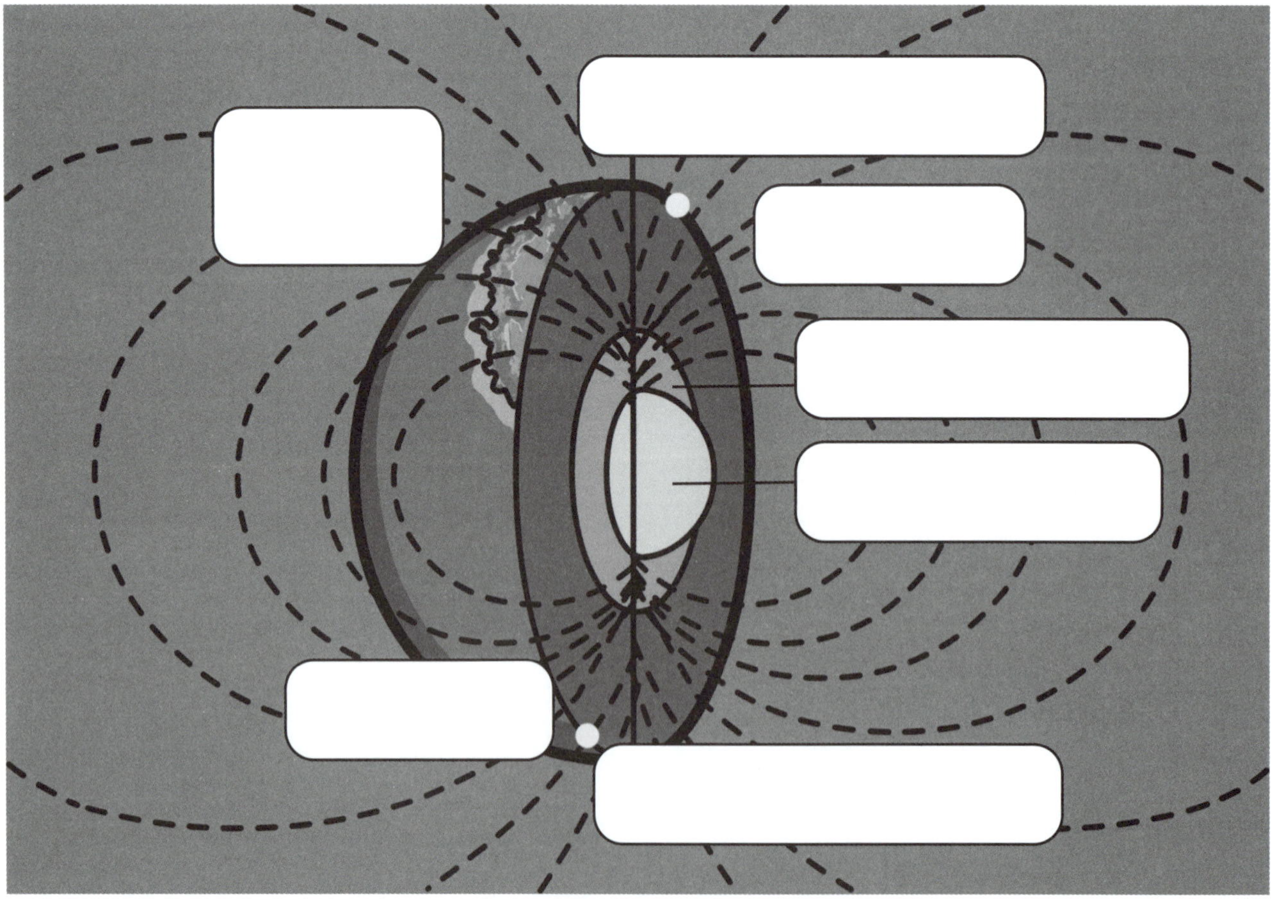

How do magnets react together?

How do magnets react together? This activity supports the investigation on page 94 of your Student Book.

How can you make your observations accurate?

You can often use measuring equipment to make measurements.

1 List all the pieces of measuring equipment that you know.

2 In this investigation, you do not need to make measurements. You need to observe. This means to look carefully at what happens.

3 How will you record your results?

Think about these questions:

How will I keep my results neat and tidy?

Will I use a table?

Will I draw a chart or graph?

4 My table, chart or graph will look like this:

Which materials are magnetic?

Magic magnets

1 You are going to move a metal object with a magnet.

   Ask an adult to help you cut a hand hole in one end of the box.

You will need: a large cardboard box, two magnets, small metal objects such as paperclips and toy cars.

2 Put your hand inside the box. Hold the magnet so that it touches the top of the inside of the box.

3 Place a paperclip or other metal object on top of the box.

4 Move the magnet around under the box.

   The magnet attracts the metal object, so the object moves with the magnet.

5 You are now going to design a race track for racing toy cars. Your design should use magnets in the same way as for the box investigation. Draw your track on the top of the box.

6 Place a metal toy car on the top of the track.

7 Move the magnet to make the car move as fast as you can.

8 Give someone else another magnet and a toy car.

9 Race the cars next to each other on the track. Who is the winner?

Floating paperclips

You are going to investigate how to make a paperclip float in the air.

You will need: a paperclip, a bar magnet, 20 cm of cotton thread or fine string, reusable adhesive or modelling clay.

Ask permission to stick some reusable adhesive or modelling clay onto a table or work surface.

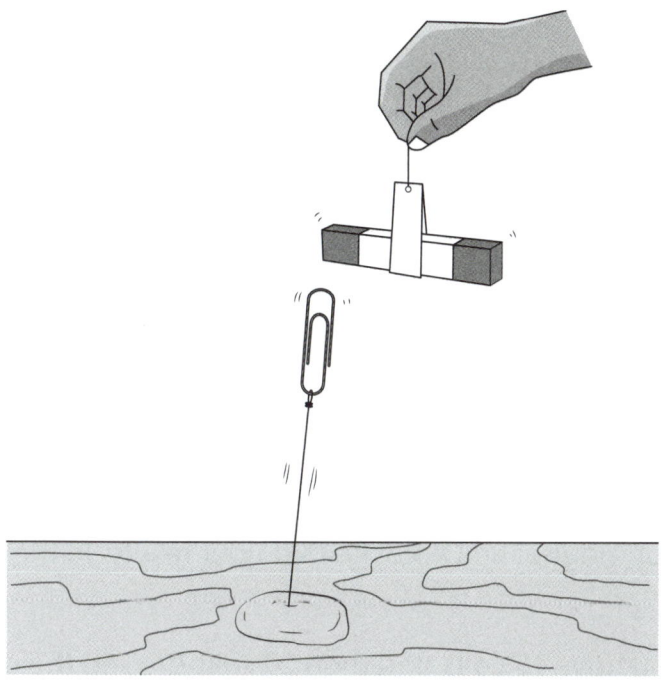

1   Tie one end of the cotton thread to the paperclip.

2   Attach the other end of the thread to a table using reusable adhesive or modelling clay.

3   Lay the thread and the paperclip on the table.

4   Hover the magnet over the paperclip.

5   Pull the magnet as high as you can away from the paperclip.

6   Keep practising until there is a gap between the magnet and the paperclip.

- The paperclip now looks as if it is floating.
- Demonstrate this to the people at home.
- Let them try to make the paperclip float.

Explain why the paperclip appears to float.

Electromagnets

Making and testing an electromagnet

1 Set up your electromagnet as shown in the diagram.

2 Coil the wire around the nail ten times.

You will need: a battery, sticky tape, a nail, insulated wire, paperclips.

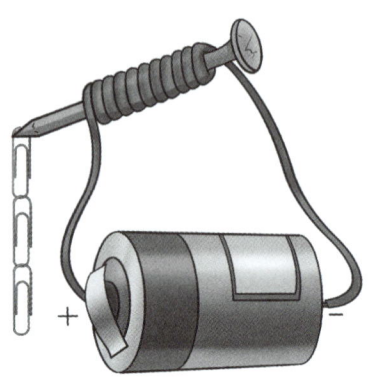

3 Add paperclips to one end of the nail. Hang them in a chain as shown.

4 Count the number of paperclips your magnet will hold. Record your results in the table below.

5 What do you think will happen if you have 15 coils and then 20 coils around the nail?

My prediction is:

_____

6 Investigate to test your prediction. Record your results.

Number of coils around the nail	Number of paperclips attracted
10	
15	
20	

7 Was your prediction correct?　　　**yes**　　　**no**

 Stretch zone

Write down a rule about the strength of an electromagnet and the number of coils.

More about electromagnets

You have made some electromagnets. The wire coils were wrapped around a nail. The nail is called the core. You will now investigate if the material used for the core is important.

You will need: a battery, sticky tape, a steel or iron nail, insulated wire, paperclips, rods of plastic, wood, cardboard and aluminium.

1 Set up your electromagnet using the iron nail with 15 coils as shown on page 98. Test to see how many paperclips it can pick up.

2 Record your results in the table below.

3 Take the nail out of the coils and see if any paperclips are attracted to the electromagnet. You are using air as the core.

4 Hold the coil over some paperclips and count how many are attracted.

5 Record your results.

6 Use different materials as the core for your electromagnet. Test the electromagnet each time, and record your results.

Material used for the core	Number of paperclips picked up
iron/steel	
air	
plastic	
cardboard	
aluminium	
wood	

7 Which material was the best for using in an electromagnet?

_____

Stretch zone

Why should you use the same battery and the same number of coils each time?

What went well

1 Think about what you have learned.

2 Talk to a friend about something that went well in this unit.

3 Tick ✓ the boxes to rate yourself.

I know that pushes and pulls are examples of forces.	That's easy. ☐ That's challenging. ☐	Pages 74–77
I understand how forces can change the shape of objects.	That's easy. ☐ That's challenging. ☐	Pages 78–79
I understand how forces can make objects start or stop moving.	That's easy. ☐ That's challenging. ☐	Pages 80–81
I know how forces can make objects move faster or slower, or change direction.	That's easy. ☐ That's challenging. ☐	Pages 82–87
I understand the forces between magnets.	That's easy. ☐ That's challenging. ☐	Pages 88–95
I know that magnets attract some metals but not others.	That's easy. ☐ That's challenging. ☐	Pages 96–99

 If you want to know more or need to check, go back to the pages in your Student Book.

Investigate like a scientist

1 Do electromagnets have poles?

   a Set up an electromagnet using a steel nail as the core. Your electromagnet should have 20 coils.

   b Use a bar magnet to find out if your electromagnet has a North-seeking and a South-seeking pole. Use the diagram to guide you.

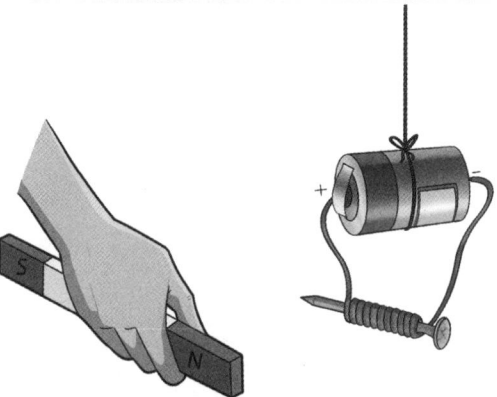

   c Is there a link between the positive and negative on the battery and any poles that you find?

2 Playing with forces

You are going to play a skittle game.

   a Place a box as shown. Fix a long stick to one side.

   b Fix a ball to the top of the stick. You can use string or elastic.

   c Stand six bottles in the box. These are your skittles.

   d Take it in turns. Swing the ball and try to knock down the skittles.

   e Count how many swings you each need. Record the results in a table.

You will need: a box, a long stick, a small ball, six bottles, string or elastic.

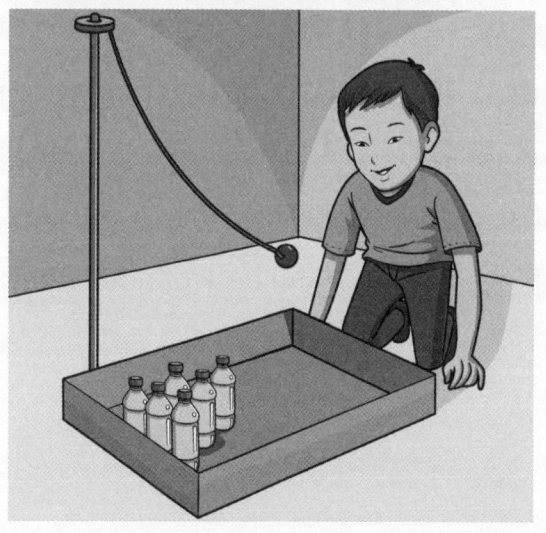

Key words

A useful way to learn key words is to try to sort out their letters when they have been jumbled up.

1 Try to sort out these key words. You can use page 102 of your Student Book to help you.

   a y h a e t l h   is sorted to make _____

   b e o b n   is sorted to make _____

   c e l s k o e n t   is sorted to make _____

   d u n i r t n t o i   is sorted to make _____

   e e c m u l s   is sorted to make _____

2 Use card to make large word cards for each of the key words.

3 Place them on your desk or your wall so you can look at them throughout the unit.

Introduction

Skeleton and muscles

 Find the words from the box in the wordsearch.

c	u	f	s	j	l	d	n	m	m
s	k	e	l	e	t	o	n	e	v
c	m	r	h	e	z	e	f	d	t
u	l	o	n	f	v	n	r	i	t
w	m	o	t	b	z	i	d	c	b
o	b	x	b	p	k	p	o	i	c
s	k	u	l	l	m	s	h	n	p
p	j	s	b	b	h	y	l	e	i
v	g	a	k	f	i	k	s	y	w
e	l	c	s	u	m	r	o	l	n

Hint: Some words might be left to right →, some might be top to bottom ↓, some might be across at an angle ↗ and some might be backwards ←!

> **bone   medicine   muscle   rib   skeleton**
> **skull   spine   symptom**

Write down the word that describes all the bones in your body.

_____

Designing a poster

You are going to design a poster to tell others that some people in the world do not have enough food.

Explain why people need food and water.

Use the words in the box.

> food give healthy help
> nutrition water

Planning

In science, it is important to plan your work before you start.

- How will you make your poster bright and interesting?
- How will you make sure the message is clear and stands out?
- Will you add pictures to help you to explain the problem?
- How will you make sure people can read the key words?
- How will you suggest that people can help?

Draw a sketch before you create your final poster.

Identifying life processes

Which life processes do the photographs show?

Use the words in the box to label the photographs.

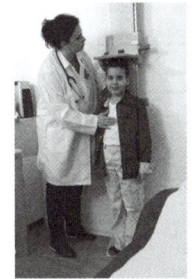

---

drinking eating growing moving reproducing

A balanced diet

How much should we eat?

The first table below shows the energy contained in each main food group.

Food group	Energy in every gram	
	Kilojoules	**Kilocalories**
fat	37	9
protein	17	4
carbohydrates	17	4
fibre	8	2

1 Which food group contains the most energy in every gram?

_____

2 Which food group contains the least energy in every gram?

_____

3 The table below shows the energy needed by different ages of people.

Age in years	Average energy needed every day in kilojoules	
	Male	**Female**
0–1	2 000	2 000
2–4	6 000	5 000
5–10	7 000	6 000
11–15	11 000	8 000
16–25	12 000	9 000
26–45	11 000	10 000
46–65	10 000	9 000
66–80	8 000	7 000

4 Young adults need more energy than older adults because _____

_____

_____ .

5 4-year-old children need more energy than babies because _____

_____

_____ .

6 15-year-old boys need more energy than 15-year-old girls because _____

_____

_____ .

Healthy eating

You are going to design a healthy breakfast. Here is a list of foods you can choose from.

> apples bananas beans bread cheese chicken chickpeas cornflakes eggs fish
> fried potatoes muesli nuts oatmeal okra onions oranges pain au chocolat
> pancakes rice squash tomatoes yoghurt

Look at the picture of a balanced food plate. Remember that if we eat the correct amount of foods from each group, we have a balanced diet.

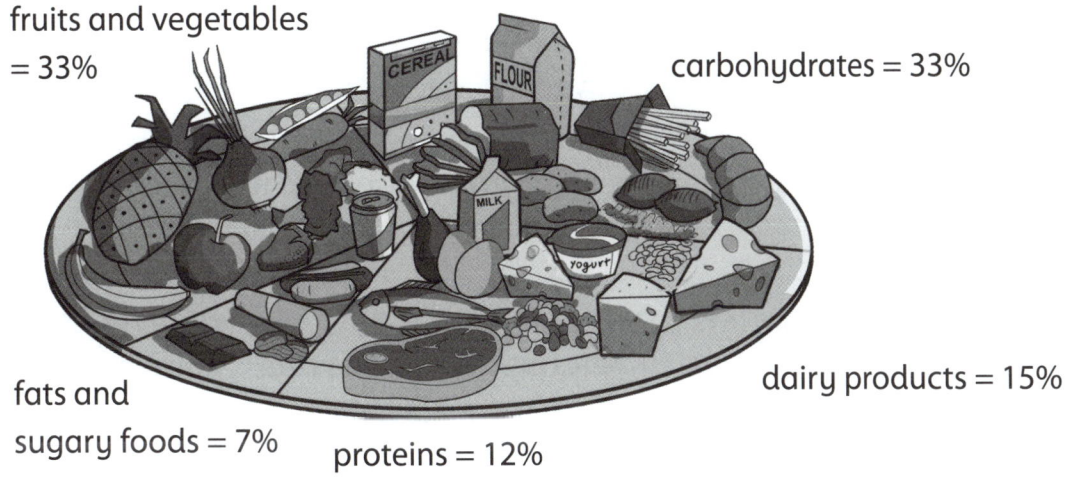

fruits and vegetables = 33%

carbohydrates = 33%

dairy products = 15%

fats and sugary foods = 7%

proteins = 12%

1 Draw a large circle for your plate. Choose some foods for a healthy breakfast. Draw the foods or write their names on your plate.

2 Think about each food carefully. Do not just choose your favourite foods. Think about making a healthy, balanced meal.

Stretch zone

Can you estimate how many kilojoules the meal will contain? Use food labels or the internet to help you with this.

Write the estimate next to your plate.

Infectious diseases

Signs and symptoms

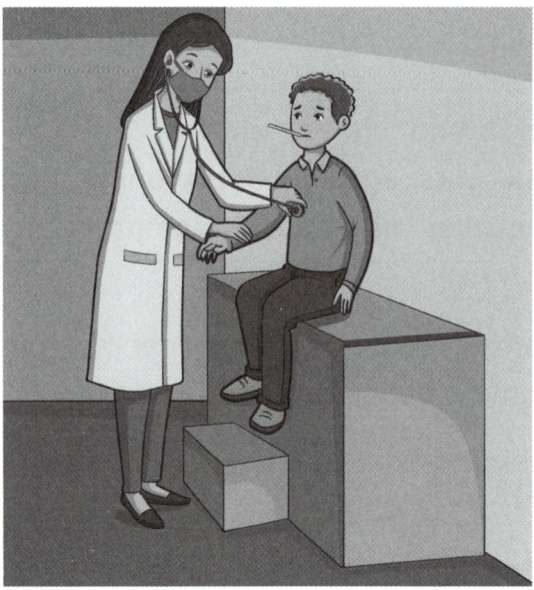

1 Read the list of signs and symptoms below. You can also look at the diagram on page 108 of your Student Book.

> coughs   feeling sick   fever   headache   high body temperature
> high heart rate   low body temperature   low heart rate   noisy breathing
> skin rash   sneezing   sore throat   stomach pain   swollen glands

2 Decide which are signs and which are symptoms.

3 Complete the table.

Signs	Symptoms

4 Write down one word which describes microorganisms that cause infectious diseases:

P_____

Lifestyle diseases

The table shows some important vitamins and minerals. It also shows some diseases that occur if a person does not eat enough of the vitamin or mineral.

Vitamin	Deficiency problems	Some sources
A	dry skin, night blindness	
C	scurvy (bleeding gums, bruising, tender skin)	
D	rickets (soft bones)	

Mineral	Deficiency problems	Some sources
calcium	weak teeth and bones	
iron	anaemia (blood deficiency)	
iodine	thyroid problems (cold, tired, overweight)	

You are going to help complete the table. You will investigate foods to find out which have a lot of vitamins and minerals.

Your teacher will provide you with some foods.

1 Look at the labels on the foods.

2 Make a list of those with high amounts of the vitamins and minerals shown in the table.

3 a Use your list to help you record your findings and complete the table.

   b Which foods contained the most minerals? _____

   c Which foods contained the most vitamins? _____

The importance of water

Looking after water

Water is important. Without water life cannot exist. It is important not to waste water.

How much water is wasted from a dripping tap?

Ask an adult to help you with this investigation.

1 Turn a tap on slightly so it drips water. Try to make it drip so that when one drip hits the sink another drip starts falling from the tap.

2 Place a measuring jug beneath the tap. Make sure the jug collects the water.

3 Predict how much water you think you will collect in one hour. _____

Remember to include the units.

4 Leave the tap dripping for one hour, then turn it off.

5 Measure how much water you have collected in one hour.

The volume of water collected in one hour is _____.

Remember to include the units.

6 Work out how much water that dripping tap will waste in one day.

Remember, there are 24 hours in 1 day. Multiply the volume collected in 1 hour by 24.

The volume of water wasted in one day will be _____.

Remember to include the units.

Stretch zone

Can you work out how much water will be wasted in one week?
Can you work out how much water will be wasted in one year?

Filtering water

It is not always easy for people to get clean water. Drinking dirty water can cause serious diseases. It is important to be able to treat water to make it cleaner. One way of doing this is to filter the water.

You will need: a plastic bottle, cotton wool, sand, pebbles, kitchen towel, jar, scissors.

You are going to filter dirty water.

1 Set up your filtering apparatus.

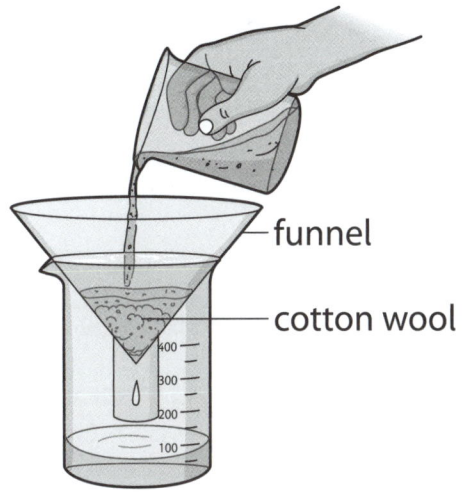

funnel

cotton wool

This is a smaller version of how water is cleaned for towns and cities. Very large filters use sand instead of cotton wool.

2 You will use some dirty water. Observe it carefully.

Write down what it looks like.

_____

3 Pour the dirty water through your filter.

How long does it take to go through the filter?

_____

4 What does the water look like after it has been through the filter?

_____

5 What does the cotton wool in the filter funnel look like?

_____

6 Where has some of the dirt gone?

_____

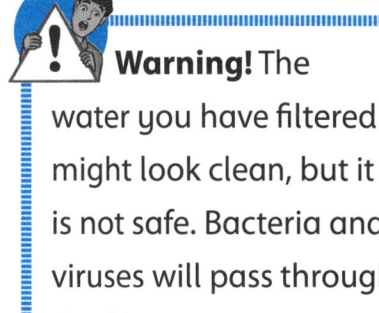

Warning! The water you have filtered might look clean, but it is not safe. Bacteria and viruses will pass through the filter.

Planning healthy meals

Food for a long walk

You are going to plan a meal for some people to take on a long walk.

Walking for a long time uses up a lot of energy.

1 Plan a lunch that the people can carry and eat during their walk.

List all the foods you advise them to take.

_____ _____ _____

_____ _____ _____

_____ _____ _____

2 Why did you select these foods?

_____

_____

3 How much water should the people take? Why?

_____

_____

Different people need different diets

Energy is measured in joules. A kilojoule is a thousand joules.

Heat energy was once measured in calories (c) or kilocalories (Kcal). How many calories are in a kilocalorie? _____

The total amount of energy a person needs during the day depends on how active they are.

Look at the table below.

Person	Kilojoules per day	Kilocalories per day
mountain climber	23 000	5 400
builder	21 000	5 000
athlete	19 000	4 500
footballer	16 000	3 800
teenage boy	12 500	3 000
office worker	11 000	2 600
teenage girl	9 500	2 300

> bananas beans bread burger carrots cereal chocolate couscous eggs
> falafel fish fizzy drink grapes ice-cream kanafeh kebab khubz lamb
> milk oranges pasta peas rice salad water yogurt

1 Write menus for the athlete's breakfast, lunch and dinner. Choose foods and drinks from the box above.

breakfast

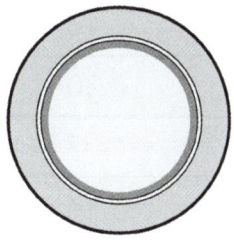
lunch

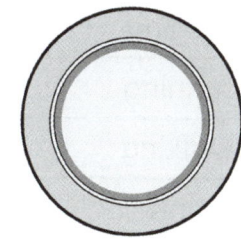
dinner

2 Think about the meals you would plan for an office worker. How will the meals be different from the athlete's meals?

_____

_____

Exercise and health

Energy and exercise

The amount of energy we need from our food depends on our age, whether we are male or female, our lifestyle, and the job we do. It also depends on how many kilograms we weigh.

A person who weighs 100 kg will burn up more energy than a person of 50 kg if they are doing the same activity.

Think back to the work you did last year on energy and exercise. This time, you are going to calculate your own energy use every hour.

1 Study the table. It shows how much energy is used up during various activities. You are given the number of kilojoules used per kilogram of weight.

2 Fill in the right-hand column to show how much energy you would use. You will need to multiply the number of kilojoules by the number of kilograms you weigh.

Activity	Kilojoules used every hour per kilogram of weight	Kilojoules you would use up every hour
basketball	20	
canoeing	12	
dancing	8	
golf	6	
jogging	18	
running	30	
walking	7	
tennis	14	

3 Explain why professional cyclists try to keep a low body mass (number of kilograms).

Stretch zone

Find out how much energy you use every hour when sleeping. What is the energy used for?

Heart rates

1   When doing exercise, your heart rate and breathing rate increase.
    Why does this happen?

    _____

    After exercise, your heart rate will slow down back to normal. It will go back to your resting heart rate.

2   Describe how you can find your resting heart rate.

    _____

    _____

    My resting heart rate is _____.

    The time your heart takes after exercise to slow down to the resting rate is called the recovery time.

3   Plan an investigation to find out your recovery time.

4   Carry out your investigation and record your findings. Use the table below.

Time after exercise (in minutes)	Heart rate (number of heart beats per minute)
0.0	
0.5	
1.0	
1.5	
2.0	
2.5	
3.0	
3.5	
4.0	

Compare your resting heart rate and recovery rate with other people in your group.

Stretch zone

Research how resting heart rate and recovery rate can be used as a measure of how fit a person is.

Remember: It is normal for different people to have different heart rates.

The human skeleton

Make a skeleton

You are going to make a model skeleton.

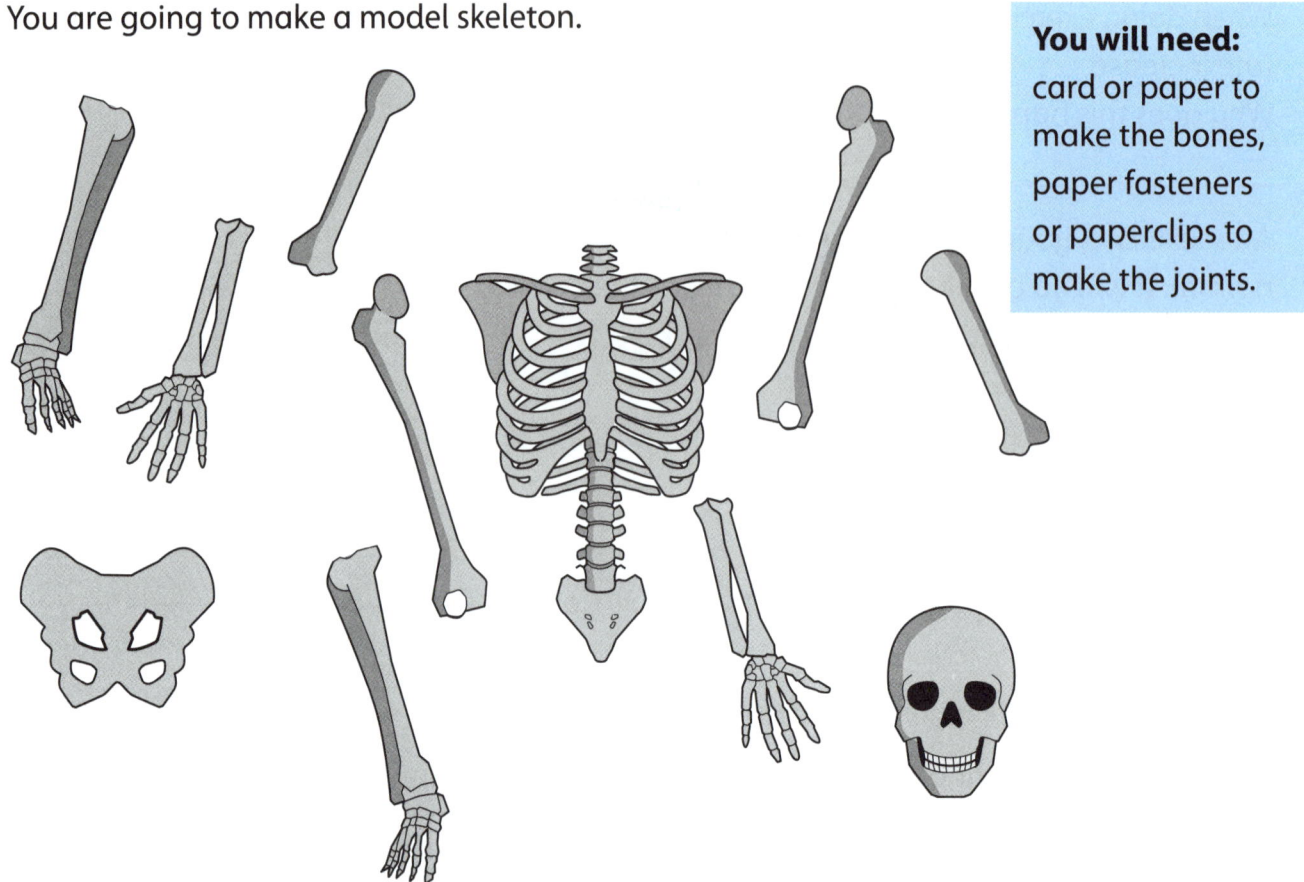

You will need: card or paper to make the bones, paper fasteners or paperclips to make the joints.

1 Copy all the bones and groups of bones onto a large piece of card or paper.

2 Put the bones in the correct places to make a human skeleton.

3 Join the bones together using paper fasteners or paperclips.

Stretch zone

What does the skull protect? _____

What do the ribs protect? _____

Why is the leg not made of a single large bone?

_____

Label the skeleton

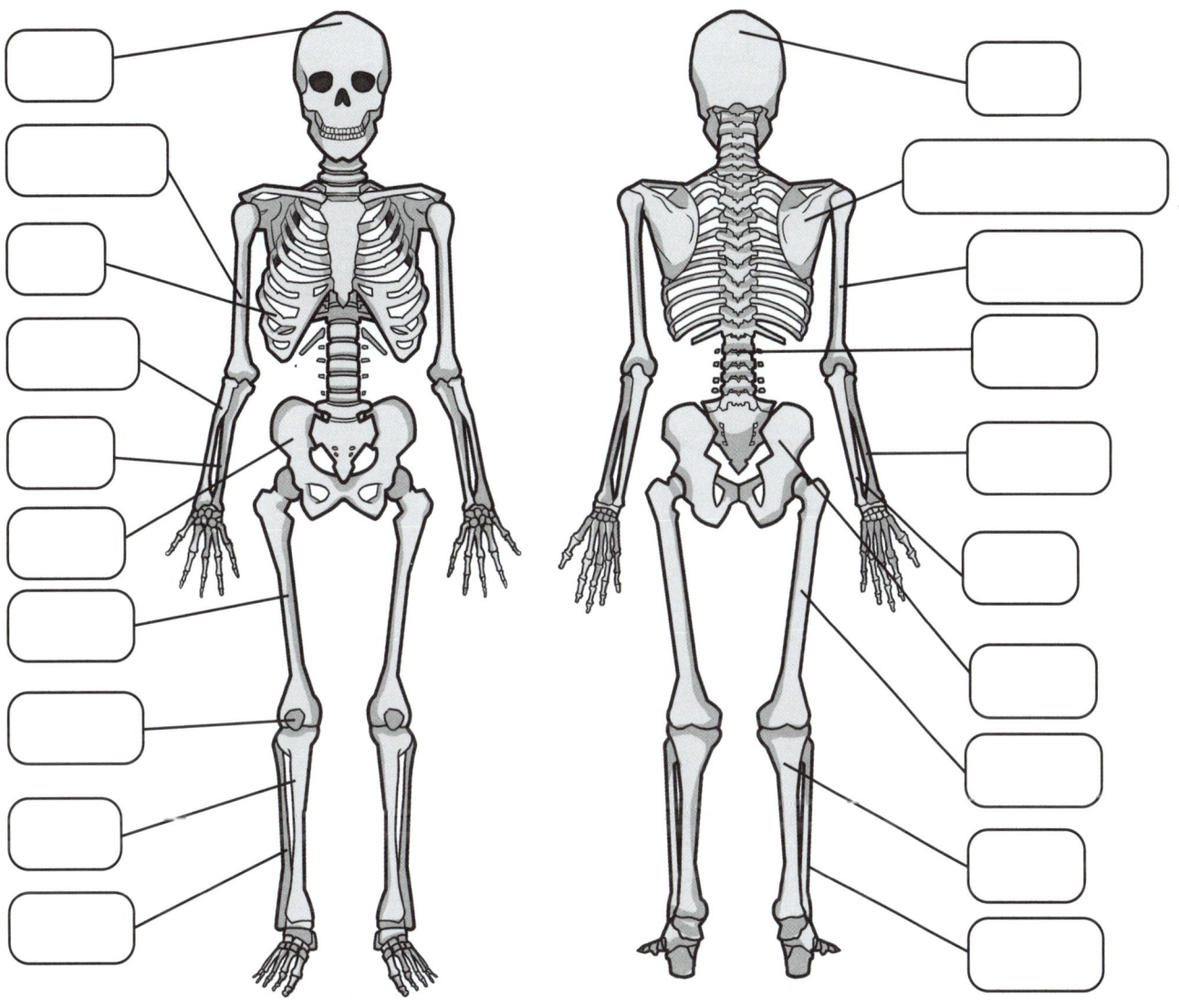

femur fibula humerus
patella pelvis radius ribs
skull tibia ulna

femur fibula humerus pelvis
radius shoulder blade skull
spine tibia ulna

1   Look at the drawings of the skeleton.

2   Label the drawings.

   ●   Use the words in the boxes to help you.

   ●   You can use page 116 of your Student Book to help you. First, try to add as many words as you can.

Comparing animal skeletons

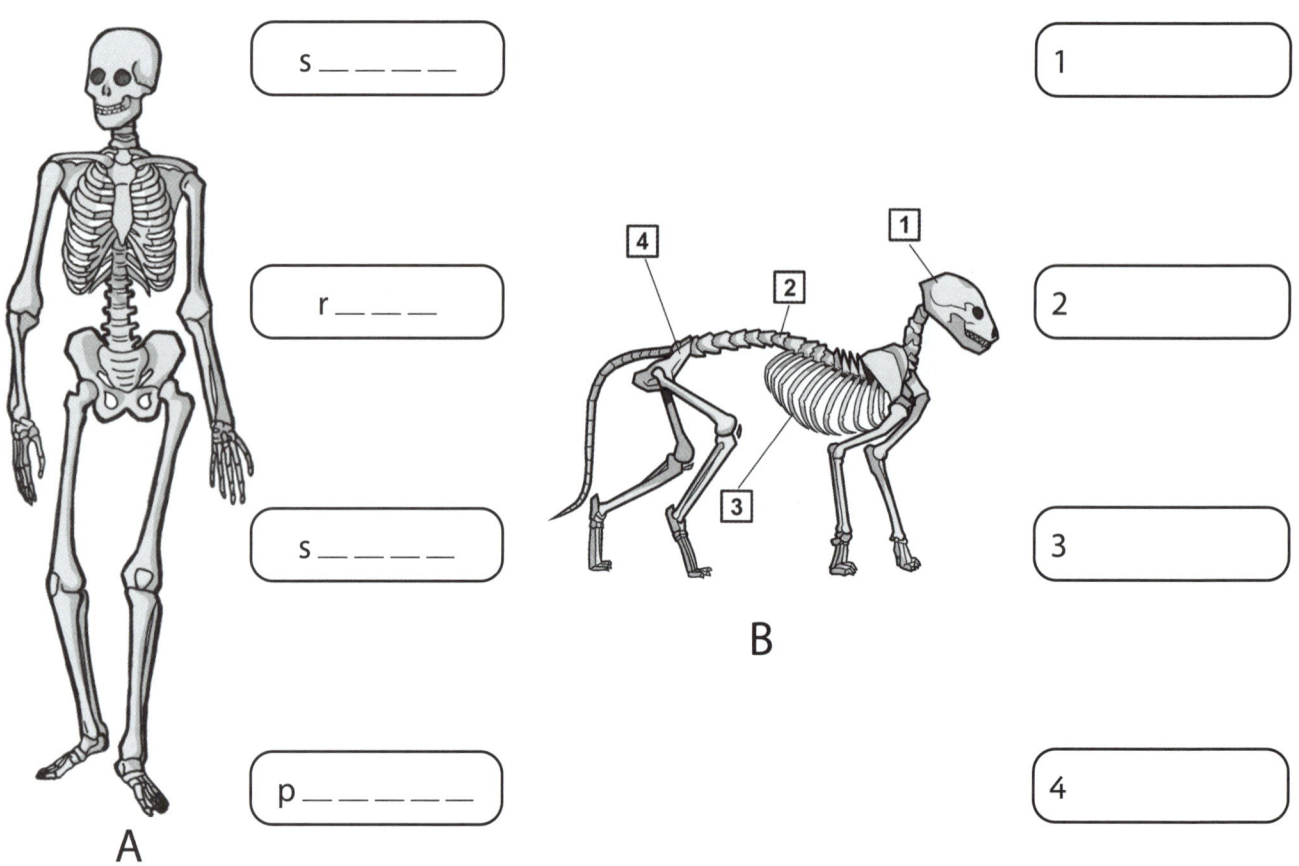

s _ _ _ _ _

r _ _ _ _

s _ _ _ _ _

p _ _ _ _ _ _

A

1

2

3

4

B

1 Look at the drawings of the animal skeletons.

- Drawing _____ shows a cat.

- Drawing _____ shows a human.

2 Label both skeletons. Draw a line to show each part of the human skeleton.

3 How are the cat and human skeletons the same?

Write down three similarities.

Matching animals to their skeletons

This activity supports the investigation on page 118 of your Student Book.

1 Which skeleton belongs to each animal? Write the letter of each animal in the boxes.

| a |  b |  c |  d |  e |  f |

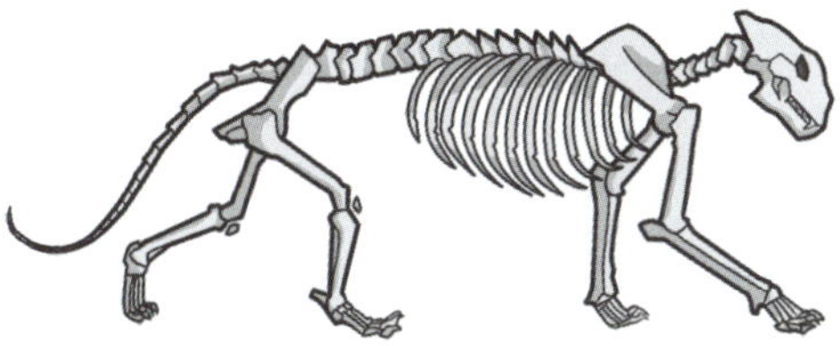

1 [ ]

2 [ ]

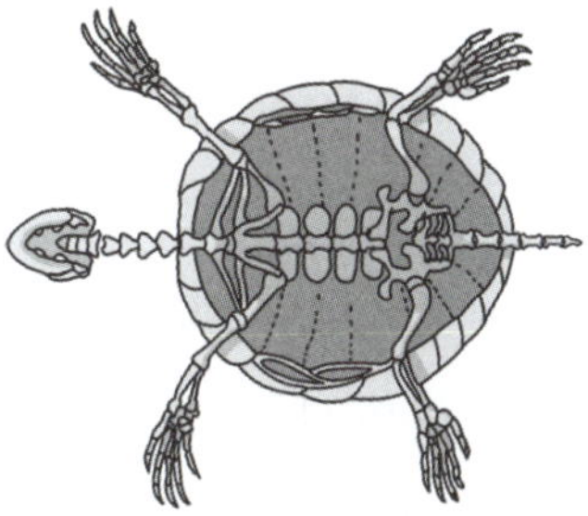

3 [ ]

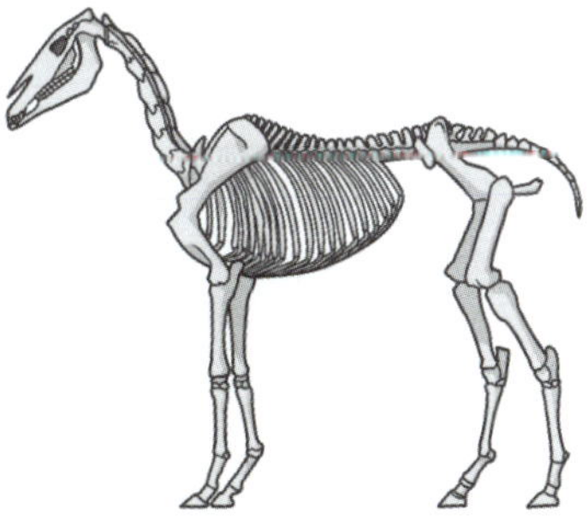

4 [ ]

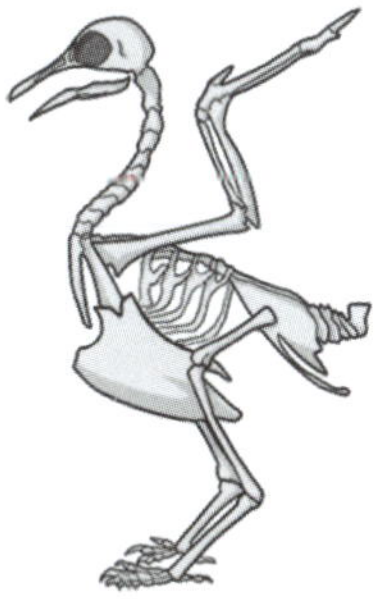

5 [ ]

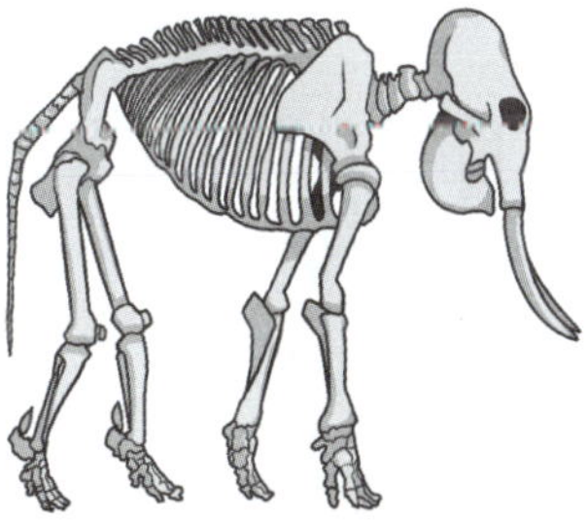

6 [ ]

2 Complete the paragraph about animal skeletons. Use the words in the box to help you.

> **differences   elephants   similar   skeleton   skull**

Many animals have a _____ inside their body. Animals have bones that are _____ to the bones in a human skeleton. These bones include the _____, the spine, the ribs and the pelvis. There are some _____ between the bones of different animals. For example, birds have wings, _____ have tusks, and frogs have long legs.

Skeletons need to grow

Growing bigger

This activity supports the investigation on pages 120–121 of your Student Book.

Do all bones grow at the same rate? You are going to carry out a survey to find out. You will measure different people's bodies.

You will need: a tape measure. If you have not got a tape measure, use string. Then use a ruler to measure the length of the string.

1 Circle your chosen phrase to complete the prediction.

I predict that bones will grow at

the same   **a different**   rate as people grow.

2 Use a tape measure to measure each person's:

- hand length
- leg length
- head circumference
- height.

You are really measuring the bones.

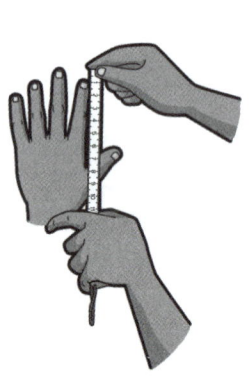

Hand length

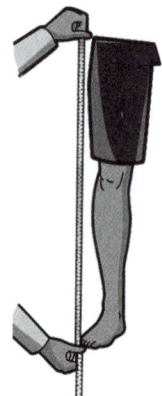

Leg length

Head circumference

3 Why is it better to measure each part more than once?

_____

4 Draw a table to record your results. Record them in order of age, starting with the youngest person.

5 Can you see any patterns in the measurements?

_____

_____

6 Do all the bones grow at the same rate? _____

Do all bones grow at the same rate?

This activity supports the investigation on pages 120–121 of your Student Book.

Use the results table below to help you record your findings for the investigation.

Measure five students.

Remember to measure each body part three times and find the average (mean).

For example, if you measure a hand three times and get 13 cm, 12 cm and 14 cm,

then the average is: $$13 + 12 + 14 = 39$$

Then divide by 3 as there are three readings: $$39 \div 3 = 13$$

So you would record an average hand size of 13 cm for that person.

Name of student	Height (cm)	Hand length (cm)	Leg length (cm)	Head circumference (cm)

Why do we need a skeleton?

Protecting your organs

Look at the diagram of the skeleton below.

One of the functions of the skeleton is to protect the organs of the body.

1 Draw and label these organs on the diagram of the skeleton.
   Make sure you draw them in the correct places.

   - heart
   - lungs
   - brain

2 Which bones protect these organs?

   Write the names of these bones as labels on the diagram.

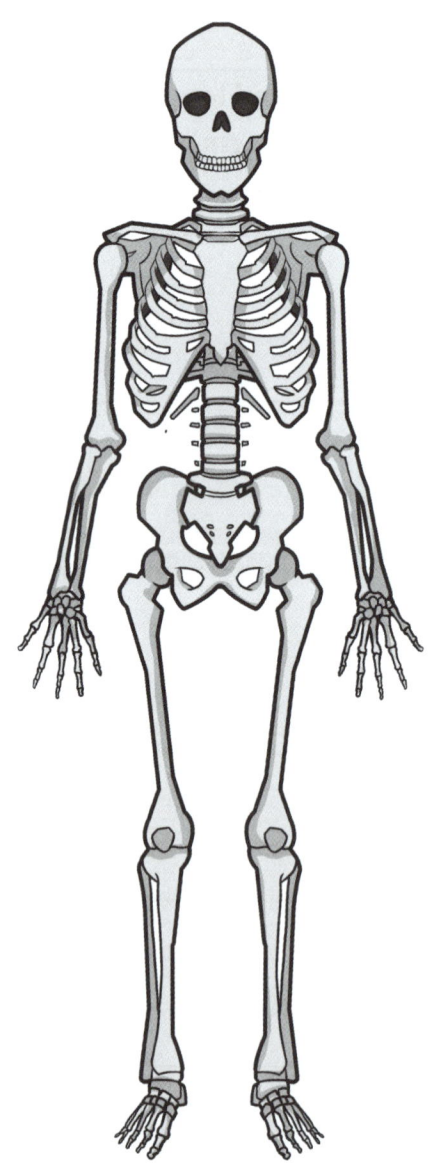

The role of the ribcage

1 Which organs does the ribcage protect? Look at the diagram and fill in the label boxes.

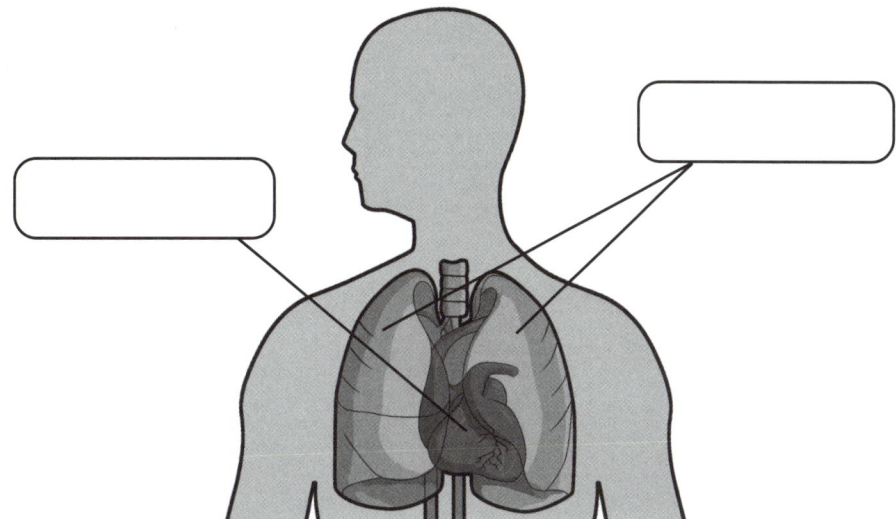

2 Complete the paragraph about the skeleton. Use the words in the box. One has been done for you.

> movement ribcage ~~skeletons~~ skull spine stand

Humans have _____skeletons_____ inside their bodies. The function of the skeleton

is to allow _____, to protect organs in the body, and to support the body.

The functions of the _____ are to protect the spinal cord, to support the

skull and to help us to _____ and sit straight. The _____ protects

the brain, and the _____ protects the heart and lungs.

Bones and no bones

Seeing through bodies

We know a lot about bones because of x-rays.

The x-rays can shine through skin and muscle and our bones show up as a photograph.

1 Using the names of the bones in the box below, label these parts of the skeleton.

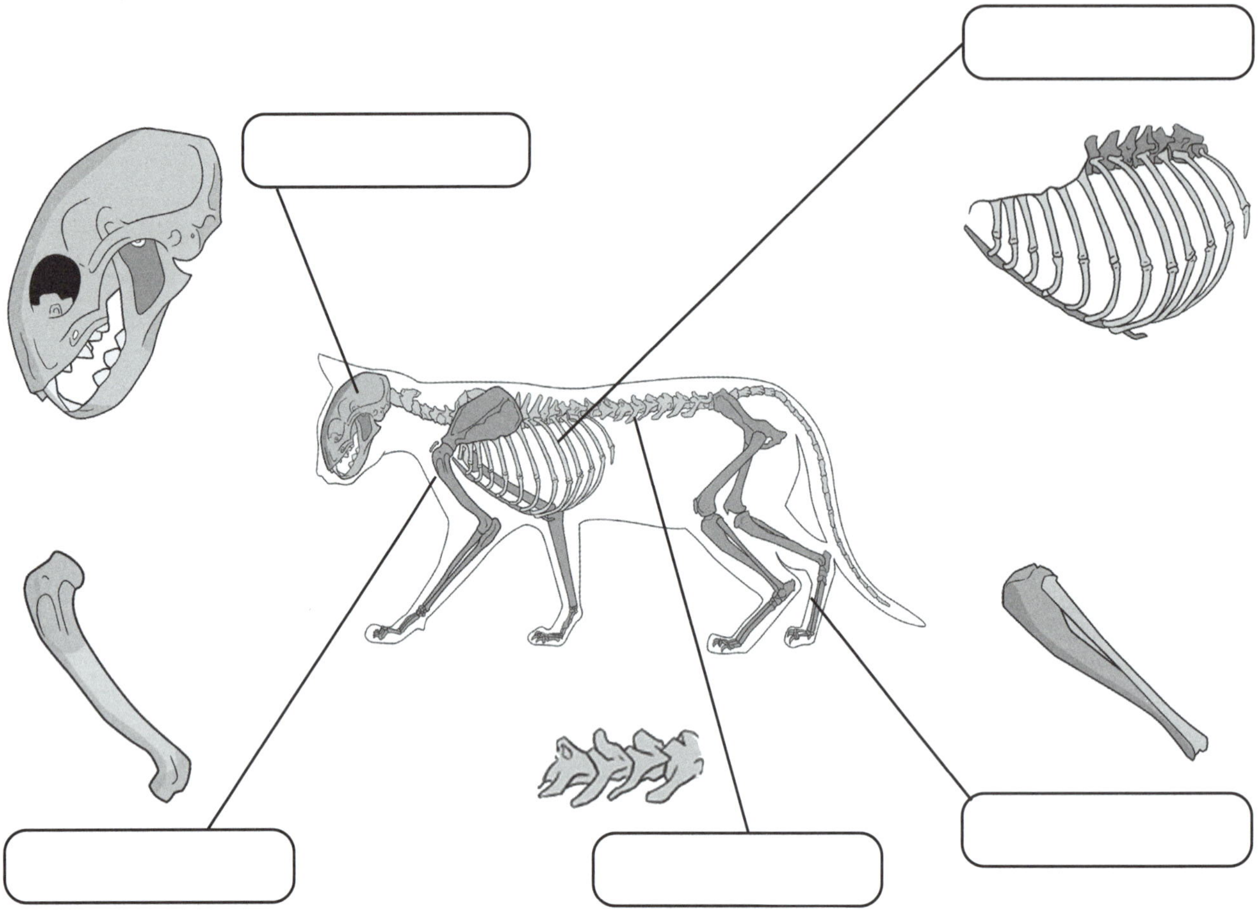

> ☐
>
> ☐
>
> ☐
>
> ☐
>
> ☐

> humerus   ribs   skull   spine   tibia and fibula

2 What animal do you think this is? _____

Stretch zone ➤

Why are x-rays important in helping people after an accident?

_____

Invertebrate survey

This activity supports the investigation on page 125 of your Student Book.

We think the best place to look for invertebrates will be:

_____ .

Warning! Never touch or go near to any animals you find. Discuss why this is important.

1 Record how many of each type you observe.

2 Use an identification key to help you name the invertebrates you observe. Record your results in a table.

A possible key for invertebrates is shown below, but you can find your own for your area.

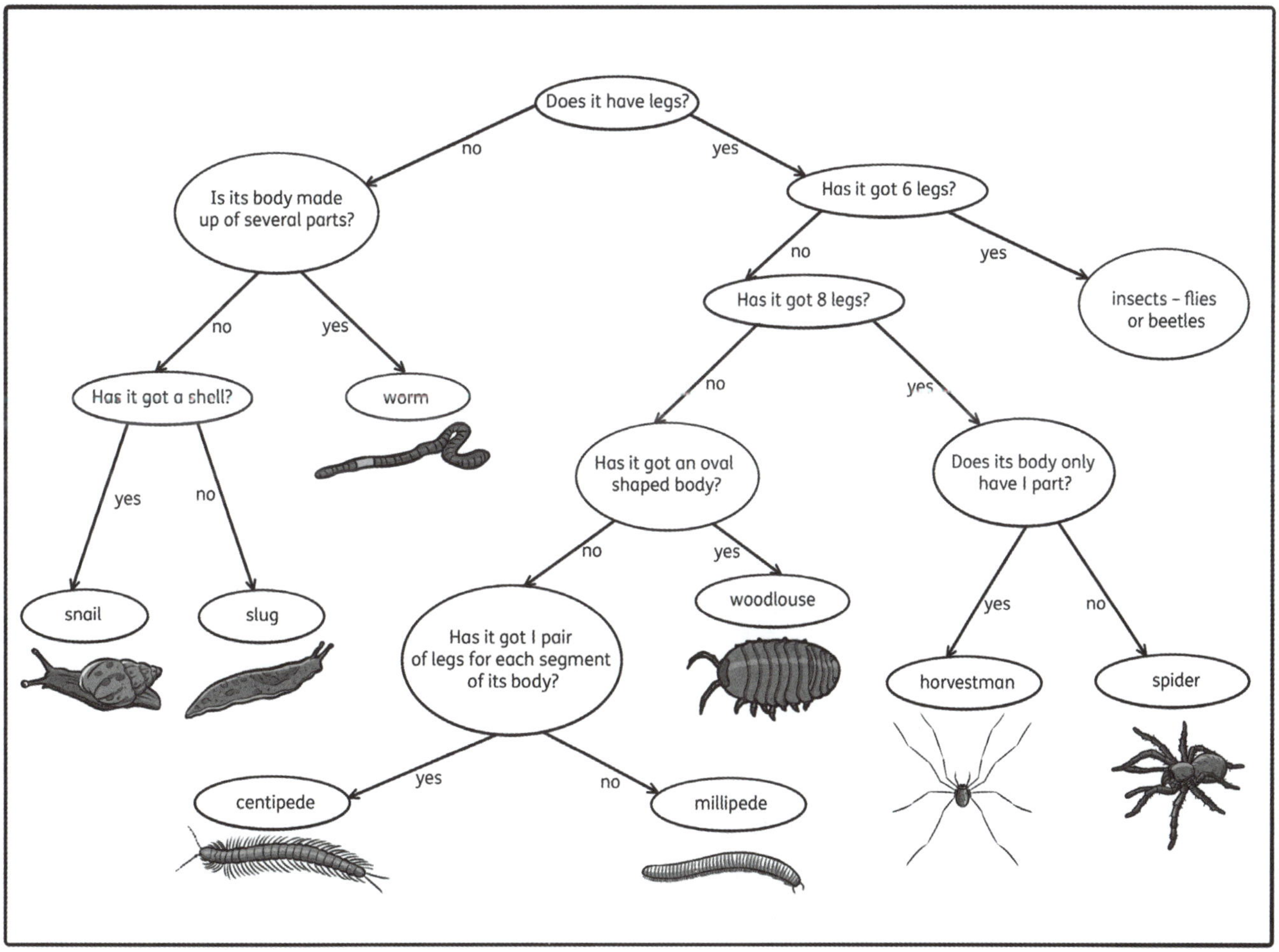

Does it have legs?

no → Is its body made up of several parts?

yes → Has it got 6 legs?

Is its body made up of several parts?
no → Has it got a shell?
yes → worm

Has it got a shell?
yes → snail
no → slug

Has it got 6 legs?
no → Has it got 8 legs?
yes → insects – flies or beetles

Has it got 8 legs?
no → Has it got an oval shaped body?
yes → Does its body only have 1 part?

Has it got an oval shaped body?
no → Has it got 1 pair of legs for each segment of its body?
yes → woodlouse

Has it got 1 pair of legs for each segment of its body?
yes → centipede
no → millipede

Does its body only have 1 part?
yes → horvestman
no → spider

Muscles and skeletons

Important muscles

Bones cannot move by themselves. They have to be pulled.

The muscles are the parts of your body that pull bones to make them move.

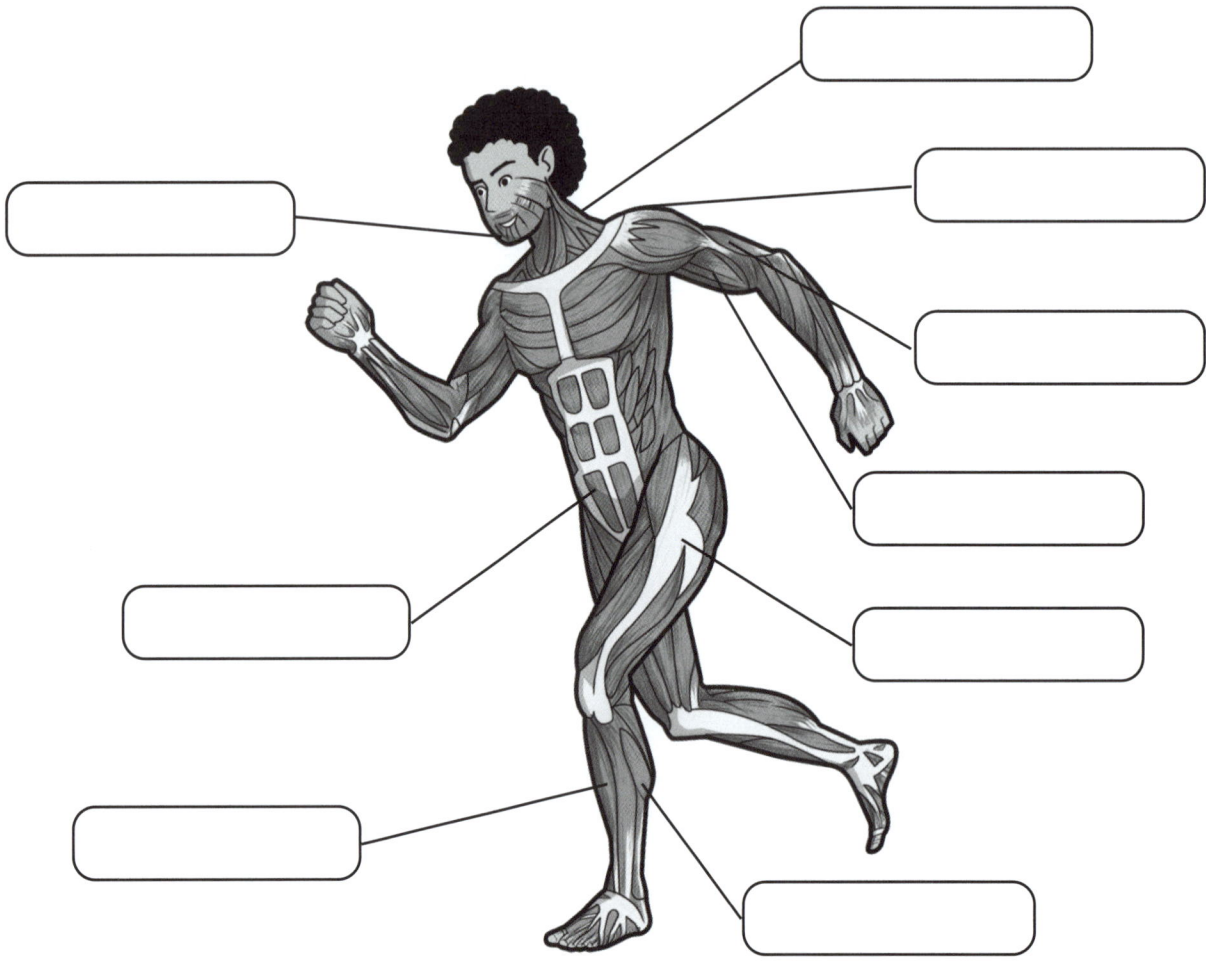

Label the muscles on the diagram. Use the names of the muscles in the box below.

> **biceps   calf muscles   jaw muscles**
>
> **neck muscles   shin muscles   shoulder muscles**
>
> **stomach muscles   thigh muscles   triceps**

Using muscles

Look at the picture of the children playing football.

1 Which muscles do you think they will be using?

2 Which muscles do you use for the following activities?

Use page 127 of your Student Book to help you.

Activity	Muscles used
walking	
talking	
eating	
running	
breathing	
picking up a pencil	
swimming	

Stretch zone

Describe two activities that use the biceps and triceps muscles.

3 Are these statements true or false? Circle the correct answer. The first one has been done for you.

Muscles are attached to bones.	**true**	**false**
Bones can move on their own.	**true**	**false**
Elbows and knees are examples of joints.	**true**	**false**
We use our jaw muscles for talking.	**true**	**false**

How muscles work together

Look, cover, label, check

1 Look at the diagram of the arm for ten seconds.

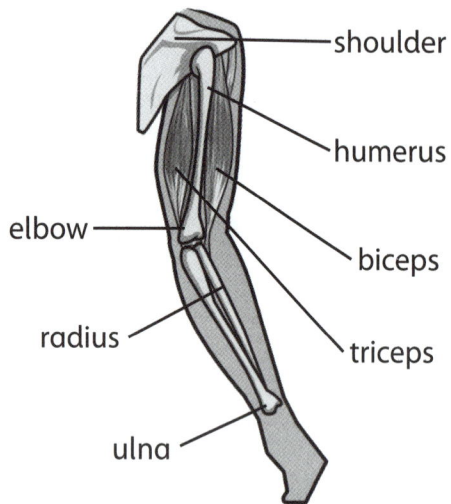

shoulder

humerus

elbow

biceps

radius

triceps

ulna

2 Now cover the diagram with a piece of paper.

3 Draw the diagram of the arm from memory. Write as many labels as you can remember.

4 If you cannot remember the diagram or labels, look at the diagram for another ten seconds and then cover it again. Keep doing this until you can draw the diagram without forgetting anything.

Hint: This is a very good way to test your memory. It is very useful before tests.

Muscle pairs

Muscles can pull but they cannot push.

Make a model arm

Your model will show the two muscles working together.

You will need:
cardboard, a paper fastener, sticky tape, two elastic bands.

1 Make the model shown in the diagram.

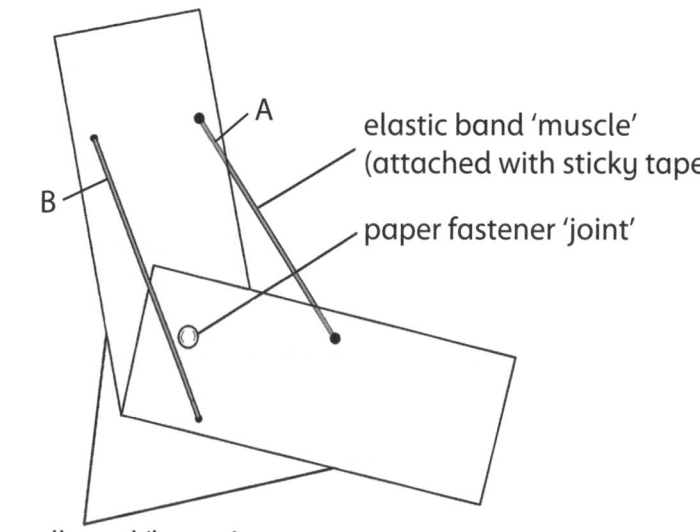

elastic band 'muscle' (attached with sticky tape)

paper fastener 'joint'

cardboard 'bones'

2 Make your model arm bend and straighten.

3 Which elastic band (muscle) pulls (contracts) to move the arm up?

_____

4 Which elastic band (muscle) pulls (contracts) to move the arm down?

_____

5 Which muscle pair do these two elastic bands represent?

A is _____.

B is _____.

Medicines

1 Look at the student's notes. She has missed some words out.

2 Use the words in the box below to help you fill in the missing words. You can use each word as many times as you need to. Some words may need to be changed slightly.

A _____ works with the body to make us feel better. Some _____ can cure _____. Some _____ can make the _____ better.

If someone gets a bacterial infection, they need to take _____. _____ kill the bacteria that make the person feel unwell. It is important to take all the _____ to make sure that all the bacteria are killed.

Some people have _____, for example hay fever. People with _____ can take a _____ called _____.

It is very important to use _____ safely. If we take too much _____ it can make us _____. _____ labels have _____ about how much _____ to take and how often to take it.

> allergies antibiotics antihistamines illnesses
> instructions medicine symptoms unwell

3 Write the name of the type of medicine that kills bacteria.

_____

Allergies

Find out about allergies

1 Ask the people at home about any allergies that they know about.

2 Tick ✓ each of the allergies in the table every time someone mentions it.

pet allergies	
skin allergies	
hay fever	
insect bites	
antibiotics	
food allergies	

3 Which allergy did the most people mention?

_____

4 Were any allergies not mentioned?

_____

 **Stretch zone**

What medicine might you take if you feel unwell with an allergy?

_____

Using medicines for a long time

Health leaflet

 This activity supports the investigation on page 132 of your Student Book.

1 You are going to design an information leaflet to help people understand asthma.

Think about the questions below. They will help you to research the illness.

How will you find out more about the illness?

_____

What causes asthma?

_____

What are the symptoms?

_____

How can the disease be prevented or treated?

_____

_____

Is there a cure for the disease?

_____

2 Now start making your leaflet.

Fold a piece of A4 paper to make three columns. You can use both sides of the paper.

Include drawings as well as writing.

Medicines crossword

Complete the crossword to see how much you remember about medicines. One has been done for you.

Clues:

1 ~~What substance do diabetics check for with a blood test kit?~~

2 Diseases caused by bacteria are treated with …

3 Asthma affects the small tubes leading to the …

4 This illness is treated by controlling insulin.

5 Hay fever is what type of health problem?

6 We can find instructions on how to use medicines on the …

7 The equipment used by asthmatics is an …

8 This illness can make us feel very hot with a runny nose.

What I have learned about health, skeletons and muscles

 What went well

1 Think about what you have learned.

2 Talk to a friend about something that went well in this unit.

3 Tick ✓ the boxes to rate yourself.

		Pages
I know life processes common to humans and animals.	That's easy. ☐ That's challenging. ☐	Pages 104–105
I know people need exercise and a varied diet to keep healthy.	That's easy. ☐ That's challenging. ☐	Pages 106–115
I know how skeletons grow, and support and protect the body.	That's easy. ☐ That's challenging. ☐	Pages 116–125
I know that muscles are attached to bones, and contract to make bones move.	That's easy. ☐ That's challenging. ☐	Pages 126–129
I can explain the role of drugs as medicines.	That's easy. ☐ That's challenging. ☐	Pages 130–133

 If you want to know more or need to check, go back to the pages in your Student Book.

Investigate like a scientist

1 Planning a healthy restaurant

You are going to plan a menu for a healthy restaurant.

a Think of a name for your restaurant.

b Plan three breakfast, three lunch and three evening meals.

- Include all the important food groups in your meals.
- Limit the amount of sugar and fats in the foods.

c Design a menu to advertise the meals in your restaurant.

d Write a short advertisement. Explain why the food in your restaurant is so healthy.

2 Making a model hand

You are going to make a model hand.

You will need: cardboard, string, drinking straws, sticky tape, scissors.

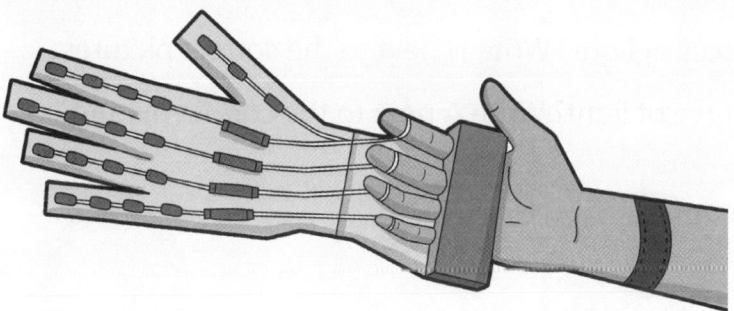

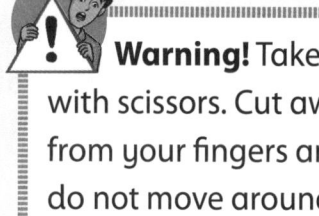

Warning! Take care with scissors. Cut away from your fingers and do not move around the room with them. Why do you think this is important?

a Cut out a hand-shaped sheet of cardboard. Make sure you cut out a long wrist with part of the arm attached to it.

b Fold the cardboard fingers where there should be joints.

c Look at the diagram. Fix string to your fingers so that when the string is pulled the fingers move. The pieces of string are acting as tendons. These are how muscles are attached to bone. Use straws to keep the string in place.

d Use your model hand to try to pick up objects.

e Make an information sheet explaining how the model hand works. Use the following words:

> bones joints muscles tendons

Quiz Yourself

How to use these questions

These quiz questions and activities are intended to encourage students to reflect on their learning and to reinforce their developing knowledge about scientific concepts in a fun way. They are flexible enough to be individual, pair or group activities. The questions can be used in a number of ways:

- Questions can be selected from this section to supplement work carried out during each module, to act as extra tasks and support for individuals, groups and whole classes. In this way, they can aid differentiation.
- Students can tackle the relevant questions at the end of each module to review learning and supplement the 'What I have learned' sections.
- Students can undertake questions at the end of a series of modules or even at the end of the year to review learning. The questions could be set in batches over a series of lessons or even taken as a small timed test – although this is not their main purpose.

1 Light and Dark

1 a Which objects are sources of light? Write S next to the correct pictures.

   b Which objects reflect light? Write R next to the correct pictures.

   c Which objects are natural sources of light? Write N next to the correct pictures.

   d Which object is an artificial source of light? Write A next to the correct picture.

2 Draw the shadow that each object will make. One has been done for you.

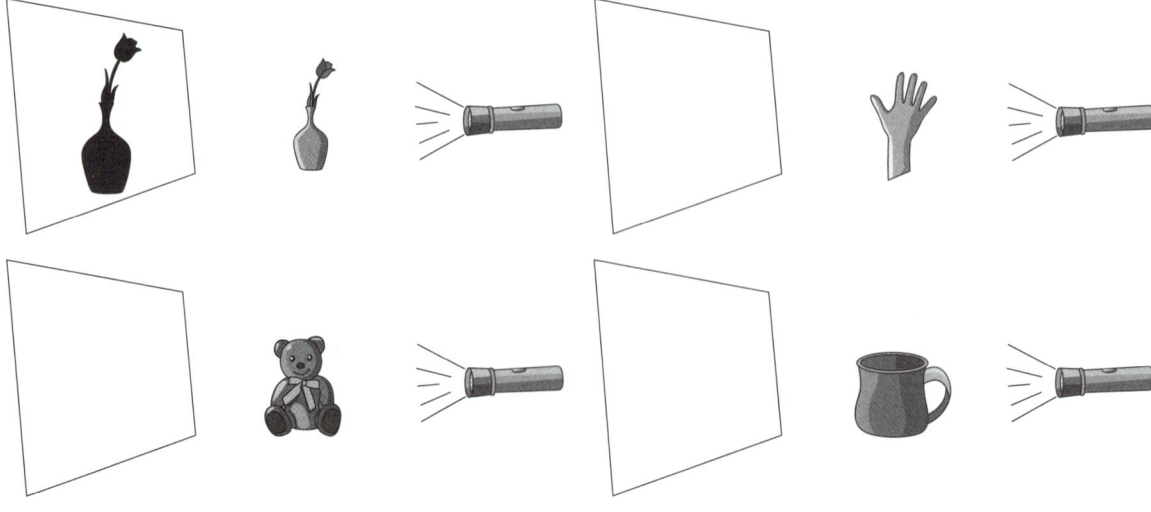

3 Look at the diagram.

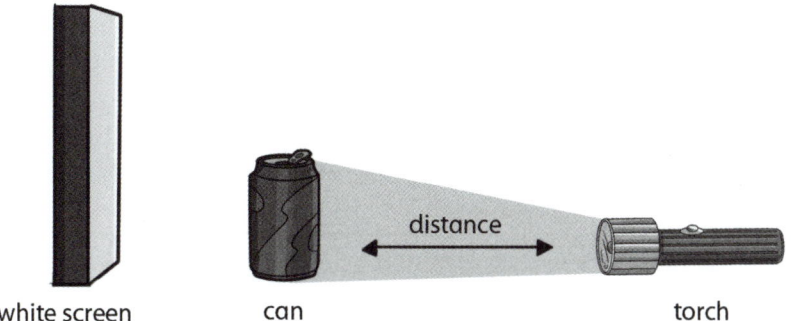

white screen   can   torch

What will happen to the size of the shadow if:

a the can is moved closer to the screen?   **It will be bigger.**   **It will be smaller.**

b the can is moved closer to the torch?   **It will be bigger.**   **It will be smaller.**

2 **Looking at Rocks and Soil**

4 How is each rock used? Draw a line to show where each piece fits.

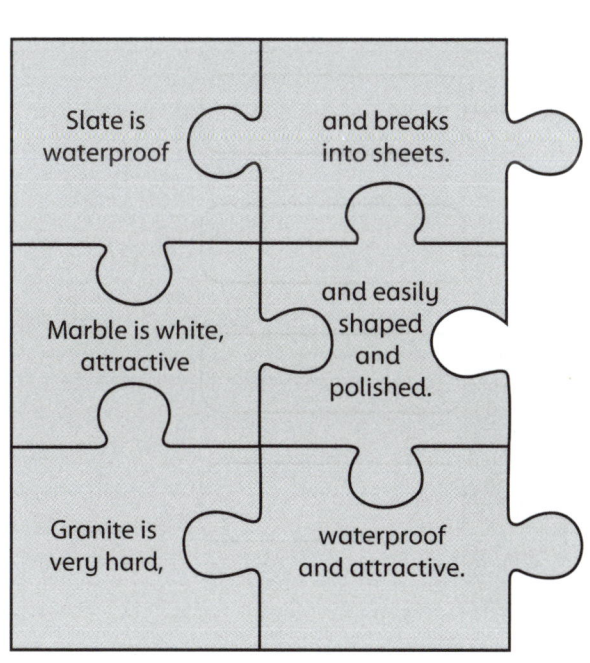

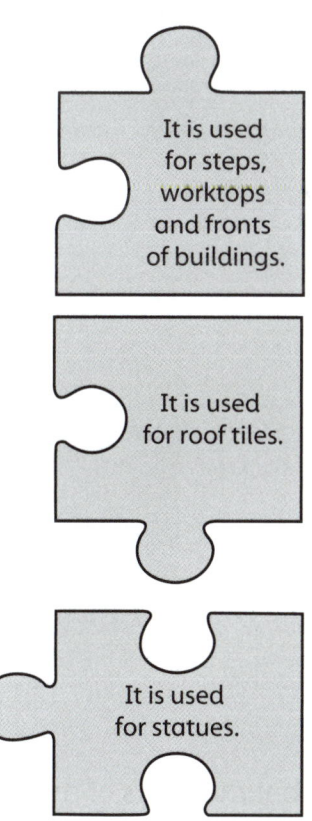

5 Look at the diagram of the layers in the soil. Write the name of each layer in the boxes. Three are done for you.

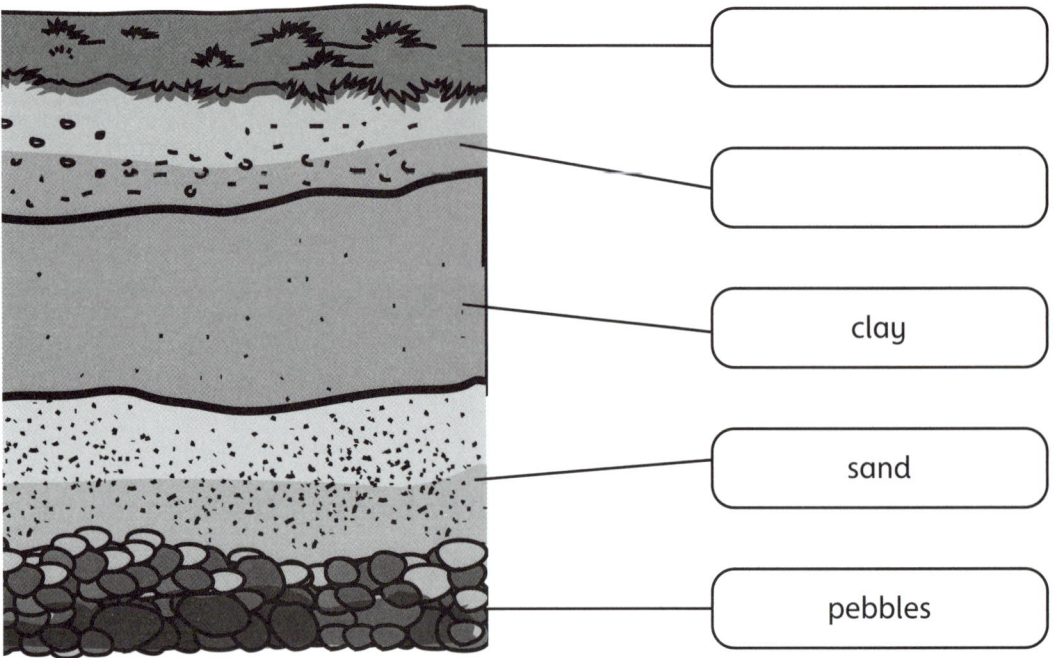

clay

sand

pebbles

3 Flowering Plants

6 Look at the drawing of a plant.

a Label the parts of the plant.

b Read the information about the different parts of a plant.
Match each of these to the correct label.

1	This supports the plant and transports food and water around it.	
2	This uses the energy from sunlight to make food.	
3	These keep the plant anchored in the soil. Some have hairs that help the plant to get water.	
4	This often has a nice smell and colour to attract insects. This is also the place where seeds are produced.	

7

a Follow the maze to help the seedling find what it needs to grow.

b What other things does the seedling need to grow?

_____

_____

4 Introducing Forces and Magnets

8 Which force is being used in each picture? Write the correct word under each picture.
   You can use the box to help you.

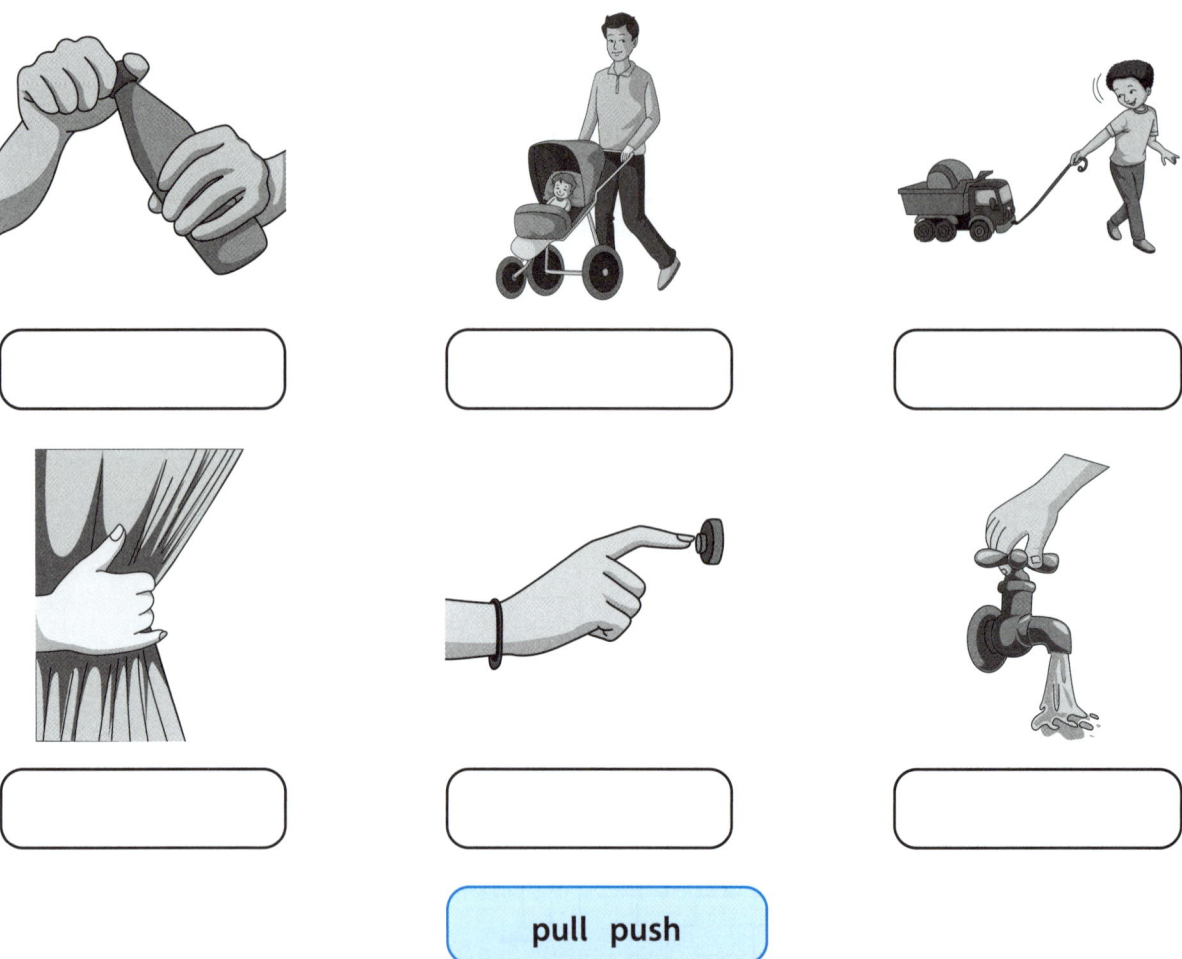

pull   push

9 Complete the crossword using the clues on page 141.

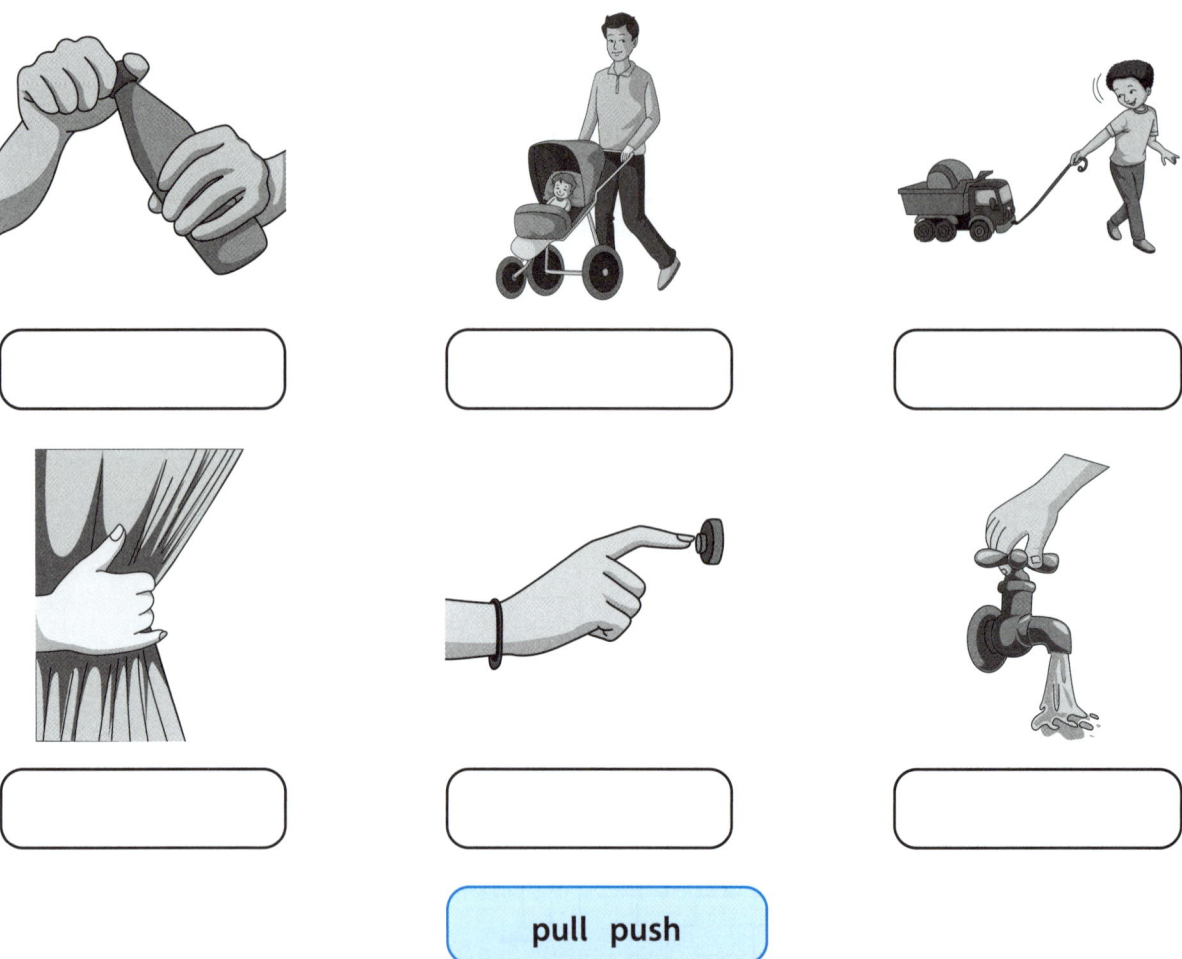

Clues:

Down

2 The North pole of a bar magnet does this to the South pole of another bar magnet.

3 We cannot see this but we can see the effect that it has. An object might change shape, speed or direction.

4 This magnetic metal can be found at the core of the Earth, as well as cobalt and nickel.

Across

1 This type of magnet is a rectangle shape and has a North and a South pole. The answer is two words.

5 The North pole of a bar magnet does this to the North pole of another bar magnet.

6 This magnet is powered by electricity.

7 This magnetic metal is a mixture of iron and carbon. Paperclips are often made from this.

5 Exploring Health, Skeletons and Muscles

10 Look at the collection of bones. Draw a line from each bone or group of bones to the correct part of the boy's body.

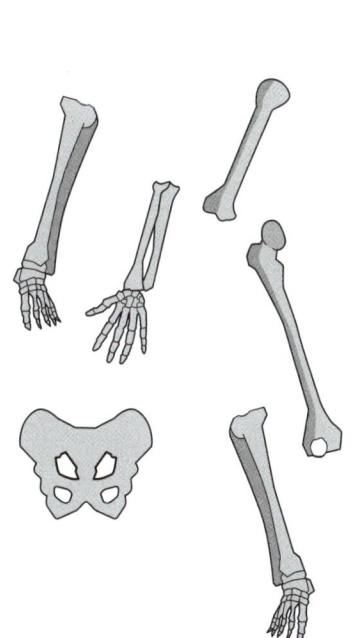

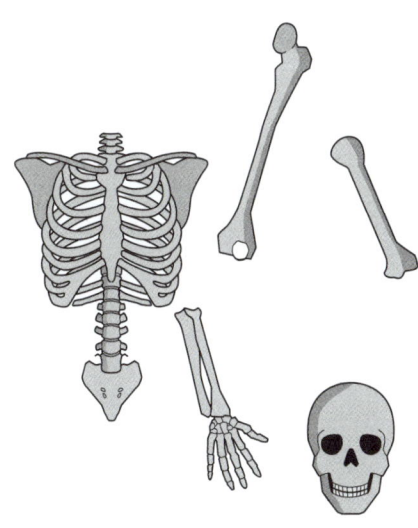

11 Find each of these words in the wordsearch.

They are all words to do with keeping healthy.

> diet drink eat exercise food growing healthy
> nutrition moving reproducing water

v	n	w	g	f	e	d	r	i	n	k
r	e	p	r	o	d	u	c	i	n	g
z	x	p	w	o	f	p	h	x	u	o
y	e	s	a	d	i	e	t	v	t	h
i	r	d	t	n	t	i	m	u	r	e
f	c	y	e	a	t	u	t	l	i	b
k	i	l	r	u	w	g	m	q	t	s
o	s	c	b	q	l	t	c	r	i	d
h	e	a	l	t	h	y	v	x	o	j
v	q	p	j	g	r	o	w	i	n	g
m	o	v	i	n	g	z	k	r	a	m